Energy in Canada

Peter R. Sinclair

ISSUES IN CANADA

OXFORD
UNIVERSITY PRESS

OXFORD
UNIVERSITY PRESS

8 Sampson Mews, Suite 204, Don Mills, Ontario M3C 0H5
www.oupcanada.com

Oxford University Press is a department of the University of Oxford.
It furthers the University's objective of excellence in research, scholarship,
and education by publishing worldwide in

Oxford   New York

Auckland   Cape Town   Dar es Salaam   Hong Kong   Karachi   Kuala Lumpur   Madrid
Melbourne   Mexico City   Nairobi   New Delhi   Shanghai   Taipei   Toronto

With offices in

Argentina   Austria   Brazil   Chile   Czech Republic   France   Greece
Guatemala   Hungary   Italy   Japan   Poland   Portugal   Singapore
South Korea   Switzerland   Thailand   Turkey   Ukraine   Vietnam

Oxford is a trade mark of Oxford University Press
in the UK and in certain other countries

Published in Canada by Oxford University Press

Library and Archives Canada Cataloguing in Publication

Sinclair, Peter R., 1947–
Energy in Canada / Peter Sinclair.

(Issues in Canada)
Includes bibliographical references and index.
ISBN 978-0-19-543386-9

1. Energy policy—Canada. 2. Energy industries—Canada.
3. Power resources—Canada. I. Title. II. Series: Issues in Canada

HD9502.C32S495 2010   333.790971   C2010-902057-X

Cover image: Shutterstock Images

Printed and bound in Canada.

1 2 3 4 — 13 12 11 10

ANCIENT FOREST™
FRIENDLY

14
trees were saved
for our forests

Preserving our environment
Oxford University Press chose to print the pages of this
book on recycled paper and saved these resources[1]:

energy	water	greenhouse gases	solid waste
4 million BTUs	24,149 L	601 kg	176 kg

Printed by **Webcom Inc.** on
Legacy TB Natural 100% post-consumer waste.

FSC

Mixed Sources
Product group from well-managed
forests, controlled sources and
recycled wood or fiber
Cert no. SW-COC-002358
www.fsc.org
© 1996 Forest Stewardship Council

[1]Estimates were made using the Environmental Defense Paper Calculator.

Contents

List of Tables

List of Figures

Abbreviations

BP	British Petroleum
CAPP	Canadian Association of Petroleum Producers
CDM	Clean Development Mechanism
EIA	Energy Information Administration
HMDC	Hibernia Management Development Corporation
IEA	International Energy Agency
IPCC	Intergovernmental Panel on Climate Change
NAFTA	North American Free Trade Agreement
NWT	Northwest Territories
OPA	Ontario Power Authority
OEB	Ontario Energy Board
OPEC	Organization of Petroleum Exporting Countries
UNFCCC	United Nations Framework Convention on Climate Change
WHO	World Health Organization

Preface

Abbreviations

A few years ago, my historian friend Sean Cadigan and I undertook a research project on the development of the oil industry offshore Newfoundland and Labrador. As this project progressed, I became increasingly interested in the global energy situation and, in particular, the energy-related issues that Canada faces.

A sociologist by training, I have also been involved with several large projects that study how human activity interconnects with the physical environment (see Ommer et al. [2007]). These "social-ecological" systems are complex, moving at an unpredictable pace and marked by unpredictable outcomes. They may change into other systems with different core characteristics; again, this process cannot be predicted. This approach is well suited to studying energy, which sits at a critical intersection of people and the physical world.

The uncertainty that surrounds complex systems does not mean that all outcomes are equally likely or that there is no value in researching interventions that might generate more desirable outcomes. To do so requires understanding the dynamics at work, particularly the interests and powers of those who most influence our contemporary industrial system. This will help to answer questions such as why has Canada been so slow to develop alternative energies or act on the issue of climate change, even when there is substantial public support for such actions Why has energy security become a major problem once again? Why has the development of oil and gas resources created so much controversy around social and environmental impacts? I will address each of these questions throughout my discussion of energy in Canada.

Thanks to Lorne Tepperman for trusting that this book would be worthwhile. For making it more accessible to you, my warm thanks to Katie Scott and Jennie Rubio, both editors at Oxford University Press. Jennie put many hours not only into refining my prose, but also raising several substantive points.

Introduction

The way we currently use energy to run our lives is unsustainable. Canada, along with the rest of the world, is rapidly facing two key issues: climate change and "peak oil" (the point at which our ability to extract petroleum cheaply and efficiently from the planet's resources starts to decline). There is little doubt that in the coming months and years, energy production and consumption will change. So will governmental policies influencing how energy is acquired and used. What are the factors that will shape these changes? How should policy evolve, and in whose interests? This book considers how these issues will affect Canada, with an eye on our relationship with the complex, increasingly globalized world beyond our borders.

A secure and affordable supply of energy underpins the economies and lifestyles of all industrialized countries. When resources become scarce and prices increase, a potentially unstable energy supply becomes a major political concern. Recall the Arab-Israeli war of 1973–74, the early years of the Iran-Iraq war (1980–88), and the unprecedented global consumption of energy in 2005–8. More complicated questions arise from these political concerns. For example, should our government take ownership of this complex problem, or rely on market-based solutions?

Another major concern is the pollution generated from our consumption of fossil fuels. This pollution is driving climate change (not to mention individual health problems). States are now coming together at the international level to consider large-scale interventions, investments, practices, and regulations. These negotiations are complicated and require compromise from all countries, rich and poor alike.

Many strategies have been proposed for addressing these problems. These strategies are interconnected, varied, and wide-ranging; they emphasize conservation, technological innovation, intensive exploration (often in fragile environments), unconventional sources (including oil sands), and substitution of alternative or renewable sources of energy for various types of fossil fuel. However, there can be contradictions. For example, coal is abundant and cheap— resolving the issue of security and scarcity. But the increased burning of coal creates more air pollution than any other major source of energy. This makes effective policy contentious and difficult to implement.

We know that we live in an increasingly globalized world. Globalization is a complex process, and not just a tendency to uniformity. At the same time, however, no society can be fully independent. What happens in one society is increasingly conditioned by what happens elsewhere. Energy is now a major concern, in Canada as well as elsewhere in the world. The scale of our usage has threatened security of supply for industrialized lifestyles and production processes. This has resulted in climate change and pressure on oil supply. The complex international network that we inhabit is approaching a tipping point. The questions of when we reach this tipping point—and what happens when we do reach it— are pressing, and not just to Canadians.

This book will explore how much global capitalism affects Canada and conditions what Canadians can do. This does not mean that we are without choices. It does, however, mean that making important choices without a full understanding of the context (particularly the US context) would be unwise at best. Chapter 1 looks at the general trends in world energy production and consumption and the possibility that annual production of oil has reached, or is nearing, its peak.

Chapter 2 looks at fuel security as a political issue that has waxed and waned in priority over the last four decades. I consider the history of state intervention on matters of supply and price, from the minimal controls of the 1960s to the National Energy Program of the 1980s. Chapter 3 examines how the oil and gas industry contributes to social inequality in Canada. Has it been a curse or a blessing for less developed areas and poorer groups? There is no clear-cut answer. This chapter evaluates the experience of Newfoundland and Labrador with offshore development and also some recent impacts in the arctic.

In recent years, global warming and climate change have become critical issues. Chapter 4 presents an overview with particular note of the changing corporate stances on global warming. I discuss the Kyoto protocol and Canada's policy up to the Copenhagen meeting in December 2009. In the months following Copenhagen, we have seen little change in Canada's policy. This leads to special consideration of oil sands and electricity, the subjects of chapters 5 and 6.

As conventional sources of oil and gas become more difficult to find and expensive to develop, and as action to counteract climate change becomes increasingly imperative, our interest grows in developing alternative and renewable sources of energy. Chapter 7 evaluates these different sources, with a focus on transportation as a key component. I discuss nuclear energy—an important source of electricity in parts of Canada for decades—before turning to renewable sources. My short conclusion summarizes the complicated issues associated with energy consumption and production in an uncertain future.

Chapter 1

Peak Oil

The recent history of energy production and consumption provides fascinating insight into how we have arrived at our current situation. This chapter considers different types of energy production before exploring the multifaceted topic of peak oil.

However one interprets peak oil, it seems clear that the world has already come to a critical moment: future oil scarcity, environmental damage, and adverse social impacts must be addressed. And most experts agree that time is running out.

The History of Global Oil Development[1]

The development of the oil industry is a history of corporate concentration and globalization. Even early in the twentieth century, it was controlled by vertically integrated, multinational companies. By the 1920s, seven companies had become dominant: Standard Oil of New Jersey, Socony (Standard Oil of New York), Socal (Standard Oil of California), Texaco, Gulf, Anglo-Persian, and Royal Dutch Shell. Some changed their names, two were taken over in the decades that followed, and several other major companies joined the survivors of the early dominant group. Earlier US anti-trust legislation did break up Standard Oil, but failed to decentralize the industry.

The "seven sisters" tried to limit their competition at the infamous 1928 meeting of company heads at Achnacarry, a castle in Scotland. The major US, British, and European companies divided the world into zones of influence with the aim of controlling production and prices. Their corporate networks intertwined to form a larger oil industrial network. Although the major companies attempted to avoid overproduction (which would have lowered their profits) through alliances, they were also forced into arrangements with states that owned the oil resources to avoid the nationalization of their investments. The companies initially dominated their agreements with developing countries, especially in the Middle East, where they successfully explored for oil.

Apart from isolated incidents of nationalization, the first real challenge to this industrial dominance was the formation of OPEC (the Organization of Petroleum Exporting Countries) in 1960. This was an attempt by the producing countries to retain more of their resource's value by forming national oil companies (for example, SaudiAramco) and negotiating more effectively with the multinationals. OPEC, which was itself split by internal rivalries, had only a moderate impact globally until the mid-1970s. In the aftermath of the Arab-Israeli war of 1973–74, OPEC was able to control production and raise prices.

Paradoxically, the higher prices of the 1970s and 1980s, coupled with concerns in industrialized societies about future supplies, spurred the exploration of areas that were only plausible to develop on the basis of those higher prices. As big discoveries were made in the North Sea and northern Alaska, these regions became major new sources of crude oil. OPEC stumbled forward, internally divided and with less overall control of prices. Production quotas and prices were set but then collapsed as individual countries responded to pressures to sell larger quantities at a lower price. In the late 1980s, the OPEC strategy of setting fixed prices was overtaken by the emergence of new benchmarks—the spot contract prices for West Texas Intermediate, Brent (North Sea), and Dubai. OPEC countries remained sufficiently important in that these spot prices were influenced by the amount that OPEC released into the market. However, as one commentator has noted, "OPEC cannot calibrate prices with any precision, nor within a set time-frame. It depends entirely on increasing or decreasing supply at the margin, and then waiting to see what happens" (Parra 2004, 322).

Most experts agree that the first Gulf War of 1991 was probably waged to ensure that Iraq could not establish control over Kuwait's enormous reserves. Saudi production countered with a short spike in prices before and during the conflict so that there was no actual shortage of oil. However, a range of factors—Saddam Hussein's survival in power in Iraq, UN sanctions, the oil for food program, and Iraq's counterstrategies—contributed to a sense of instability in this area that ultimately became a central feature of the world's oil production. The invasion and removal of Saddam Hussein in 2002 by a US-led consortium, under the flimsy justification that Hussein was assembling "weapons of mass destruction," only served to make matters worse. Opposition to the US invasion, the ongoing Israeli-Palestinian conflict, and the struggle between radical and moderate Islamic factions endangered the security of imports to the world's largest consuming countries.

We will see below how the failure to discover major oil reserves elsewhere, combined with unstable conditions in the Middle East, created a situation approaching crisis point in world energy supplies by 2007–8. Meanwhile, the national oil companies extended their operations and controlled the supply of some 80 percent of world oil reserves. The multinationals remained powerful through their control of downstream operations, but were constrained by agreements that were less favourable to them as well as by their lack of independent access to the largest sources of oil.

World Energy Trends

The recent growth in world energy consumption is disquieting. Over the last four decades, economic expansion and individual tastes have resulted in a threefold increase in the use of primary energy. In recent times, industrializing countries—led by China, Brazil, and India—have vastly increased their consumption. In fact, by 2007 the OECD group of industrial countries (including Canada and the US) accounted for barely half of the global total. Figure 1-1 shows that much of the increase in the last decade is due to the emerging economies; OECD countries have used more energy, but have grown at a slower rate. The European Union's demand for energy rose only slightly in comparison since 1980. The former Soviet Union even shows signs of de-industrialization.

Figures 1-2 and 1-3 indicate that Canada's energy consumption grew more quickly than that of the US and with less severe downturns in 1973–74 and 1979–83. This is perhaps the result of the US context. There has been greater pressure to conserve in the US, which increasingly depends on imports. Much of these imports are from Canada, its single most important source.

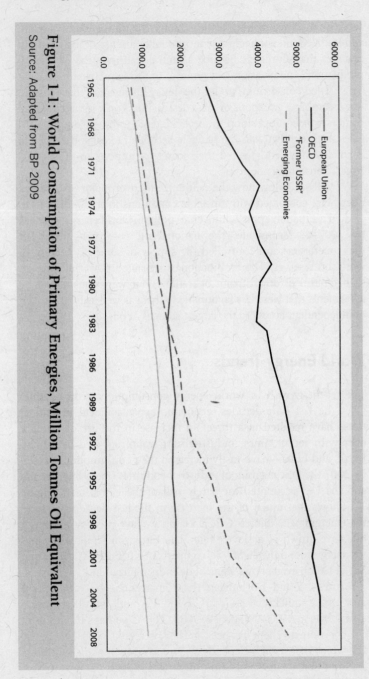

Figure 1-1: World Consumption of Primary Energies, Million Tonnes Oil Equivalent
Source: Adapted from BP 2009

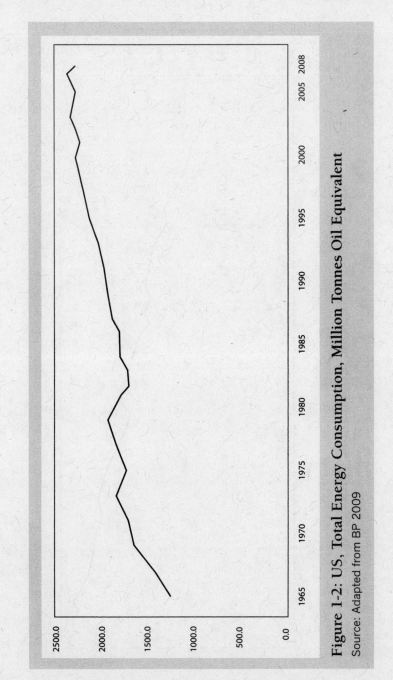

Figure 1-2: US, Total Energy Consumption, Million Tonnes Oil Equivalent

Source: Adapted from BP 2009

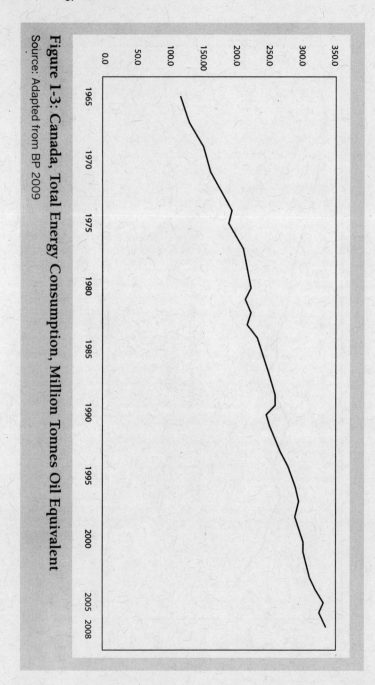

Figure 1-3: Canada, Total Energy Consumption, Million Tonnes Oil Equivalent

Source: Adapted from BP 2009

Oil and gas dominated commercial energy sources in 2008, with oil alone accounting for 36 percent, closely followed by coal at 29 percent. All fossil fuels contributed 89 percent of what was consumed in this year (Figure 1-4).

On a global scale, fossil fuels are clearly vital to maintain our standard of living. Nevertheless, there is considerable variation in lifestyle across societies. Figure 1-5 compares energy usage in Japan (an industrial society with few natural resources), China (the newly industrializing powerhouse), the US (the world's largest consuming society), and Canada (resource-rich relative to its internal demand for energy). Of these four countries, Japan is the most oil-dependent and China the least, but China relies enormously on coal to fuel its growth. Natural gas is relatively important in North America, and Canada is slightly ahead of China in the share of energy that comes from hydro developments. About 32 percent of Canadian energy is *not* based on fossil fuel, but the US figure is only about 11 percent. The addition of renewable energies would not, at present, change these figures substantially. For the world as a whole, renewable energies—including wind,

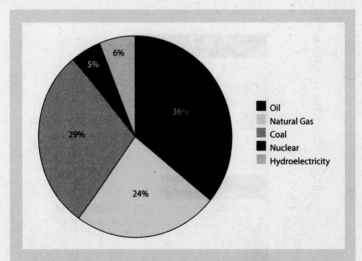

Figure 1-4: Percent of Total World Consumption, Various Energies, 2008

Source: Adapted from BP 2009

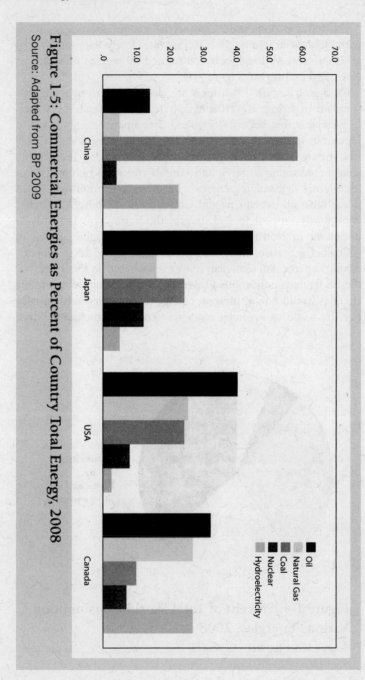

Figure 1-5: Commercial Energies as Percent of Country Total Energy, 2008
Source: Adapted from BP 2009

solar, and geothermal—still accounted for less than one percent of consumption in 2008 (IEA 2009).

Oil production

Oil is the world's largest single fuel source, and it is by far the most important for transportation.

Oil production has risen dramatically since 1965. As reflected in Figure 1-6, the overall growth was interrupted to an extent in 1974–75 and more substantially in the second oil crisis that followed the Iranian revolution of 1979 and the Iran-Iraq war. From the mid-1980s, production increased again with only minor interruptions, despite major crises in the Middle East. For the world as a whole, oil supply reached its highest point of 86,653,000 barrels daily in July 2008. "Oil supply" is defined by the Energy Information Administration as the production of crude oil (including oil sands and lease condensate), natural gas plant liquids, and other liquids, and refinery processing gain or loss. Crude oil also peaked in July 2008 at 74,798,000 barrels daily before dropping to 71,929,000 by March 2009 (*International Petroleum Monthly*, May 2009).

The overall expansion of oil production masks several important changes in the political and geographical structure of the industry. Allowing for substantial yearly fluctuations, it is clear that production in OPEC reached a peak in the 1970s and then plummeted in the 1980s (the time of the Iran-Iraq war) to a low from which it gradually recovered. However, it did not exceed the level of the 1970s until 2003–4 (Figure 1-7). OPEC was in no position to determine prices consistently. By the early 1980s, oil from non-OPEC sources (excluding Russia) accounted for more of the world total and by 2005 still exceeded OPEC's contribution. The former Soviet Union was already a major international source of oil in 1965, but exports declined following the post-Communist upheaval in the years after 1990. By the early twenty-first century, Russian production was again on the rise and Russia, indeed, was second only to Saudi Arabia as a producing country.

There are several trends in international oil production that have socio-political implications. The US dependence on imported oil has strongly coloured its foreign policy. This dependence has likely contributed far more than the administration acknowledges to its two wars against Iraq. What appears critical to US policy is the combination of the country's long-term decline of oil production

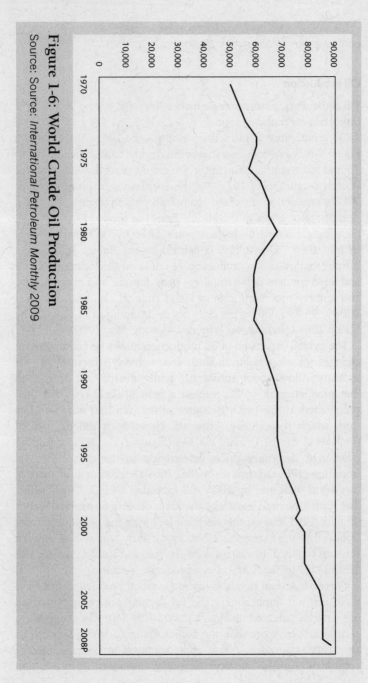

Figure 1-6: World Crude Oil Production

Source: Source: *International Petroleum Monthly* 2009

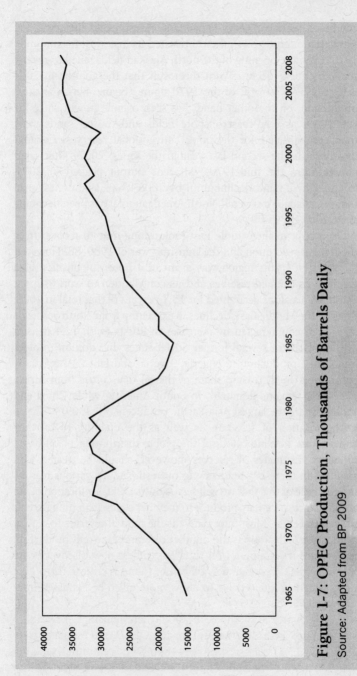

Figure 1-7: OPEC Production, Thousands of Barrels Daily

Source: Adapted from BP 2009

combined with its energy-dependent industry and lifestyles. The extraction of crude oil in the US reached its historical maximum in 1970. Even the opening of the north Alaskan fields failed to reverse the subsequent decline, with the result that the volume in 2007 was only 59.5 percent of the 1970 figure (Figure 1-8). Canadian production, on the other hand, has seen overall growth since the late 1980s as east coast offshore fields and the oil sands more than compensated for declining conventional reserves. Canada's maximum is the second last year in the series: 2007. After rapid growth from the mid-1970s, Mexico's output peaked in 2004, followed by a rapid decline until by 2008 it was slightly less than that of Canada. The overall North American total has been declining since 2003 (BP 2009).

Oil output in the Middle East took a long time to recover from the Iranian revolution and the Iran-Iraq war of 1980–88. However, by the 1990s, this region was again in a powerful position as a result of its claimed reserves and its contribution to world supply. Saudi Arabia alone accounted for 13.1 percent of that total in 2008. Outside the Middle East, declines in extraction from North Sea fields have been counteracted by increases in Russia (with 12.4 percent of the total) and several former Soviet states. In addition to older established production in Nigeria, Algeria, and Libya, Angola now contributes an increasing share of the oil that comes from Africa. Venezuela remains dominant in South America, while China and Indonesia are the largest Asia-Pacific producers (BP 2009).

The location of reserves, as well as the quantity of oil they contain, are key indicators of the global distribution of oil-based influence. Estimates of reserves, however, should be treated with skepticism: there is a tendency to overestimate for various reasons (an example of this that we will see below is OPEC's allocation of the amount each state may produce to member states according to their proven reserves). Most informative is the ratio of reserves to annual production. This ratio is the number of years at current production levels until the supply of oil runs out (unless new discoveries are made). Whether enough new oil can be found is critical; this is true even if technologies evolve to allow more oil to be extracted from known fields.

For the world as a whole, in 2008, there remained 42 years of supply. For North America and for the OECD, the estimates were 14.8 and 13.2 years respectively. For OPEC countries, another 71.1 years were available. Moreover, OPEC's reserves amounted to

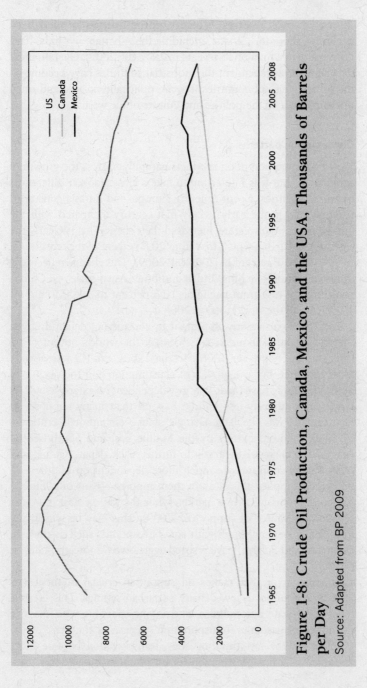

Figure 1-8: Crude Oil Production, Canada, Mexico, and the USA, Thousands of Barrels per Day

Source: Adapted from BP 2009

76 percent of all reserves. Whereas proved reserves for the US are declining, those for Canada, including the Albertan oil sands, have been rising to a total of 28.6 years in 2007 (BP 2009). Overall, these data show how dependent the industrial countries have become on the oil resources of countries that are politically volatile and often unsympathetic to the politics and culture of the west.

Consumption patterns

World consumption of oil products naturally mirrors the growth of production, but it is interesting to take a closer look at differences between countries. North America, Europe, and Eurasia consumed *relatively* less in the early twenty-first century compared with the 1960s, while other regions increased their share. This is clearly the case in Asia Pacific, which rose from 20.8 percent of the world total in 1990 to 30.1 percent by 2008 (BP 2009). This change reflects the industrialization of China, India, and other Asian societies. China's consumption rose from negligible (0.3 percent in 1965) to almost 10 percent of the world total in 2008.

The US is particularly important in considering demand, given that it consumes about 22.5 percent of the world's output of oil (this was the figure for 2008). In the 1960s, the US was already a net importer, but the growth of consumption and the decline of domestic supply have made it largely dependent on external sources. However, it imports less Middle East oil than many suppose. In fact, Canada was the largest single source of imported crude oil in 2008, followed closely by the Middle East and South/Central America with West Africa a little further back. Japan, China, and other Pacific countries are much more dependent on Middle East oil, which supplies almost all of their imports. Meanwhile, Europe has become Russia's prime market, while the Middle East is a close second as a source of supply (BP 2009). This may be why many European states (although with some exceptions, such as the UK, Denmark, and Poland) have resisted aggressive US foreign policy in the Middle East.

Another factor that makes oil supply so crucial in the US is housing organized in low density, suburban settings. This lifestyle relies on private automobiles, and is only possible with massive quantities of gasoline. Transportation, in fact, is the crux of the problem. In the last three decades, transportation became more dominant as a final destination for oil products (non-energy uses,

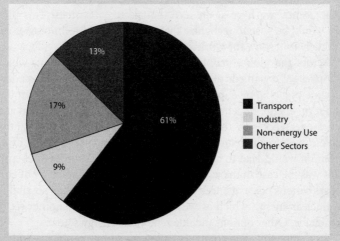

Figure 1-9: Uses of Oil, World 2007
Source: Source: Adapted from IEA 2009
Note: Other sectors include agriculture, commercial and public services, and residential and non-specified sectors.

especially plastics and chemicals, also increased). By 2007, around the globe, more than 61.2 percent of oil was used in transportation. And transportation, much more than power generation, will bear the brunt of any shortfall in the supply of crude oil (Figure 1-9).

Oil as a source of electricity production is falling. In 1973, 24.7 percent of electricity was fuelled by oil. But by 2007, this share had dropped to 5.6 percent as natural gas, hydro, and nuclear expanded (IEA 2009). These changes leave electricity generation far less dependent on oil than in the past.

When Will Production Reach its Maximum Level?

Even in the early years of commercial oil production in the US, there were periodic concerns about supply. These were soon countered by new discoveries, technological innovations, and increased output. In 2008, more oil was produced globally than ever before, but recently we have again seen the spectre of scarcity. This time we would be wise to recognize that new fields and technologies are unlikely to resolve the world's growing need for oil.

Across the globe, people are heavily dependent on oil, which supplies 42.6 percent of total energy (IEA 2009). Even a gradual decline of global oil supply will have far-reaching effects, given our modern, energy-dependent lifestyles. Massive social disruption and wars over access to the remaining oilfields are clear possibilities unless we can dramatically reduce our dependence. Moving away from oil would lead to an increased use of other sources of energy; this was the case in the early 1980s. Coal is one possibility, but coal poses serious consequences for global warming. Another possibility is nuclear energy, which as we will see below, poses another set of problems. Other options include renewable energy sources.

Any argument that points to long-term supply problems will be troubling to producers and consumers of oil. And long-term supply is the basis of peak oil analysis. The concept of peak oil dates back to 1956, with the initial work of M. King Hubbert (Deffeyes 2005). In 1956, Hubbert estimated that oil production in the lower 48 states of the US would peak around 1970. It did. Thereafter, he estimated, decline would inevitably follow. US production figures indicate a second lower peak following the discovery of the giant oil field in Prudhoe Bay, Alaska.

Hubbert observed that the volume of discovered oil followed a bell-shaped curve. Annual reserves would grow slowly; the rate of growth would increase before attaining a peak. After this peak, annual decline would begin to follow a pattern similar to the rise. This prediction was accurate, with production duplicating Hubbert's described curve, albeit about 20 years later. The gap between the estimated amount that has been discovered and the cumulative amount produced from a given field constitutes the reserve for future production. So knowing the rate of discovery enables us to predict with some accuracy the potential for future extraction, and experts believe that there will be fewer discoveries since most of the major oil resources have now been found. Deffeyes (2005) argues that this simple explanation remains accurate and has not changed substantially despite new technologies, deep-water drilling, and higher prices.

Hubbert and others—often geologists formerly employed by oil companies—began to calculate the future of global oil production.

This is a more daunting task. Future demand must be estimated, geo-political factors interfere with production, and estimates of reserves from key producing areas are unreliable. And there are yet more complicating factors. First, there is no consensus on how to count reserves. Another complication is that technological advances have made it possible to extract more oil from a given field than in the past. Even so, production is exceeding new discoveries. And as time passes, these discoveries are becoming, on average, smaller. In spite of a good deal of investment in exploration, no super-giant fields have been found since 1968 (Hirsch et al. 2005). Robelius (2007) states that both the oldest major producing area (the US) and the newest (the North Sea) are clearly in decline. It also appears unlikely that geologists have missed new producing areas that are sufficiently large and accessible to compensate for those now in decline, or that must eventually reach the stage where they begin to decline.

The peak oil perspective has been attracting ever more attention. In 2005, the BBC reported widespread interest in peak oil analysis. The concept received influential support from a French government investigation that predicted the peak in 2013. A cautiously worded report sponsored by the US government (Hirsch et al. 2005) favours 2016 as the likely peak rather than the more optimistic assessment of the US Geological Survey, which selected 2036 (Deffeyes 2005). In 2005, the Swedish government was the first government formally to accept the validity of the theory (Olofsson 2005). Those who are skeptical of peak oil argue that high prices will generate a successful round of new exploration and the development of sites that were previously too expensive. Peak oil analysts counter that this will simply deplete the world's (admittedly unknown) supply even faster. This dispute is almost inevitable given the many unknown factors.

It is impossible to pinpoint the total amount of recoverable oil. Reserves are considered proven if there is a 90 percent or greater probability of eventual recovery; probable reserves are those assumed to have at least a 50 percent chance of recovery (Robelius 2007). For any given field, the size estimate should become more accurate as development takes place. What can be recovered is also influenced by technological change. For all these reasons, reserve growth is normal in established fields.

Evaluating reserves is further complicated by the lack of access to data on the part of independent evaluators. Both countries and

The Timing of Peak Oil

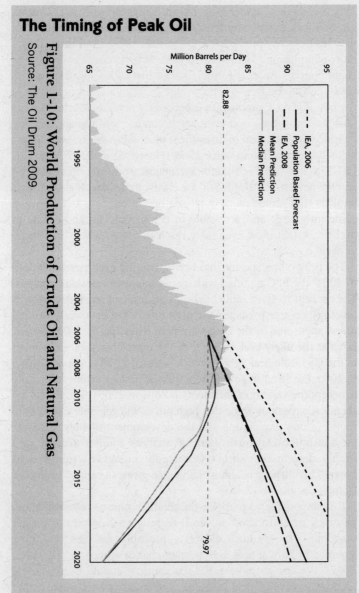

Figure 1-10: World Production of Crude Oil and Natural Gas
Source: The Oil Drum 2009

Peak oil analysts predict that there will come a point when total world oil production will reach a maximum. This is based on what can be taken from known sources operating at full capacity.

Figure 1-10 shows world production of crude oil plus natural gas liquids. The medium peak forecast is based on 15 models that predict the peak before 2020. Most models predict the peak between 2008 and 2010 in contrast to the International Energy Agency's (IEA) model of growth beyond 2020. One of the best known advocates of the peak oil approach is Matthew Simmons. Drawing on US government data sources, in 2007 Simmons claimed that the peak had already been reached—in May 2005.

As we near the end of 2007, May 2005 is still the magical "moment in time" when global crude oil peaked at 74.3 million barrels a day. Some miracle series of new oil fields could suddenly be found and quickly brought on to production, but the more time that passes, the less likely this is (Simmons 2007, 2–3).

Although this claim was rebuffed by slightly higher production in the first half of 2008, the world may be experiencing a period during which production forms an undulating plateau.

Another commentator used a different approach, focusing on the 507 known giant fields that contain about 65 percent of the world's oil reserves. Robelius (2007) selected 312 fields that had daily production of at least 0.1 million barrels per day and that accounted for 61 percent of total world production. He argues convincingly that the giant fields dominate world production and that their rate of discovery has slowed dramatically. Key factors in the future are deepwater fields and oil sands. Deepwater fields are likely to achieve maximum production in 2012 and decline quickly afterwards (Robelius 2007). Alberta's oil sands—assuming that certain technical problems can be solved—should rise to a maximum level of about 6 million barrels per day around 2040, while in Venezuela it may be possible to achieve 2.4 million barrels per day of heavy oil by 2025. Robelius expects the global peak to arrive between 2008 (in his most pessimistic scenario) to 2018 (in his most optimistic); all this takes into account planned developments, expected discoveries, successful technologies, and the annual rate of change in demand.

companies may exaggerate reserves for their own interests, with the clearest illustration, noted above, being the sudden increase reported by some OPEC countries in the 1980s. At this time, no large new discoveries were reported. However, production quotas for countries depended partly on the size of their reserves of oil. These reserves were also reported to be stable in some cases through the 1990s, despite large-scale extraction and few discoveries, or even none at all. This means that many countries were overstating

their oil reserves. Indeed, 25 countries did not change their reserve
estimates between 1995 and 2005 (Robelius 2007). From this it is
reasonable to assume that publicly stated reserves are optimistic.

The debate has become increasingly heated in recent years. This
is most likely because peak oil analysts favour public intervention
in order to plan effectively for the end of "easy oil." By contrast,
opponents of peak oil typically embrace an unregulated market
perspective; they reject suggestions that the market has failed to
produce public good or that it must now be constrained. But the
fact remains that even those who are skeptical of peak oil are in
favour of developing resources in politically stable areas. That alone
provides a strong bargaining chip to Canada.

What do global oil companies say?

Oil company managers are clearly aware that they are mining a
non-renewable resource. At the same time, they must also maintain
share value. This would not be easy if investors become convinced
that the source of profit is disappearing or becoming uneconomical
to extract. How, then, do oil company managers cope with the
prediction that oil production will peak in the next few years? No
spokesperson of any major company acknowledges that global
production of oil and gas will begin to decline by 2015. Officials of
Total SA and Chevron are the most pessimistic, but the positions of
others are unclear. Yet the difficulties experienced by companies in
recent years suggest that they are in a highly problematic situation.

Most major companies are producing considerably more each
year than they replace with reserves, and recent investment in
exploration has often failed to justify the expense (Boxell 2004).
The difficulty of maintaining production even with high prices is
partly due to reduced corporate ownership of resources in many
developing countries; in many instances, national oil companies
have taken over their own country's resources. At the same time,
it is more expensive and more difficult to find large quantities of
new oil. This becomes clear from just looking at the reports filed by
individual companies.

In general, the companies emphasize their enormous profits and
efficient use of capital rather than the threat of a declining crude
oil base. ExxonMobil takes an optimistic position. The company
does not suggest that oil is unlimited, but in 2009 its outlook report
anticipated an annual increase of 0.9 percent in oil consumption

Can the Major Oil Companies Maintain Production?

> In 2004, Shell acknowledged that it had overstated its reserves by 20 percent. In 2006, Shell stated that it was replacing only 70 to 80 percent of its production, while claiming that new fields would prevent any decline of actual production (ASPO 2006). Before the global recession, in the first quarter of 2008, crude oil production declined by 6 percent, while capital investment in exploration and production had fallen in 2007 to $13.7 billion from $15.7 billion a year earlier (Shell 2007; 2008). Overall, Shell anticipated increased total oil production as a result of its oil sands investment.

> In 2006 and 2007, BP reported declining production, but also reported that reserves continued to be replaced. However, reserves that BP controlled saw decline, and the ratio of new reserves to production was only 44 percent (BP 2007).

> Excluding oil sands, Chevron's net proved reserves of oil and gas declined by 8 percent in 2007, while net production was almost stable (Chevron 2007).

> In the first quarter of 2008, ConocoPhillips increased output slightly compared with one year earlier, but its global crude oil production fell 12 percent in 2007 from the previous year.

> Also in 2007, ConocoPhillips' proved reserves dropped by 5.4 percent, due partially to expropriation of its operation in Venezuela (ConocoPhillips 2007).

> Total's production of oil and gas liquids was down 2.6 percent in the first quarter of 2008 from the previous year. In 2007, production showed no change from 2006, but proved reserves of liquids fell by 10.7 percent (Total SA 2008).

> For 2008, ExxonMobil (2009), the world's largest producer and the one most focused on oil and gas, reported a year-on-year decline in liquids of 6 percent, while the figure for 2007 was 2.4 percent below the previous year.

The case of ExxonMobil is particularly interesting: although the company's production has declined over several years, it was able to replace its production with new reserves (as it has done consistently in recent years).

Yet, this is still misleading: the total of proved reserves of liquids declined steadily for five years—amounting to 13.9 percent since 2003—before increasing in 2009. The company masks the decline in crude oil by referring to the expansion of reserves of oil and natural gas combined and expressed in oil-equivalent barrels. In addition, heavy oil and oil sands are important parts of this total. Considering only crude oil, in 2008 Exxon announced that it did not expect to expand production by 2012 (Jubak 2008). Moreover, the company is replacing oil with less valuable natural gas (Levine 2009). In the same publication as these figures are found, ExxonMobil (2009) estimates an increase in annual oil consumption at the world level at a rate of 0.9 percent until 2030. This giant company appears unable itself to maintain production levels, a fact that should trouble its shareholders.

from 2005 to 2030, most of which was expected to come from conventional sources (ExxonMobil 2009). Chief executive Rex Tillerson claimed that reserves were ample, stating in 2008: "I know production is not peaking" (Tillerson 2008). He estimated that less than one-third of the world's total supply of oil had been consumed—but this was based on data that most commentators now consider to be a seriously inflated estimate. The casual reader is left with the impression that peak oil theory is invalid, and that much more oil remains to be extracted in the future than in the past. However, even if oil can be pumped for decades to come, the concern is that the annual demand will soon exceed the annual supply.

By 2004, BP still did not acknowledge any short- or medium-term problem, as indicated in this statement by Lord Browne, the company's chief executive: "At current levels of consumption, there are sufficient reserves to meet oil demand for some 40 years and to meet natural gas demand for well over 60 years" (BP 2004, 1). Two years later, Browne added (in response to being asked if he felt peak oil theory was "panic-mongering"):

We don't have to be worried. There are still sufficient reserves out there. Technological advances enable us to pump far more oil from a field than in the past. We used to recover about 20 to 30 percent, now it's about 40 to 45 percent, and there is no good reason to assume that we shouldn't be able to achieve 50 or 60 percent (Browne in Follath and Jung 2006, 2).

Browne reiterated his confidence in new technologies and new discoveries in Russia and West Africa, as well as the development of the oil sands in Canada.

But there are others in BP who do not stand behind this reassuring commentary. For example, one official estimated total recoverable oil at 2.4 trillion barrels with a peak that might come as early as 2010. This commentator noted that there was division in the company, with economists being more optimistic than geologists (Orange 2004). Economists expect high prices to generate greater investments in exploration and technology. Geologists, by contrast, do not see major new discoveries taking place no matter how hard they look.

Chevron, by 2006, had still not clearly stated an opinion on the future of oil, but did promote discussion on energy issues through its interactive website (www.willyoujoinus.com). Their site opened with the statement: "One thing is clear: the era of easy energy is over." Chevron has not offered a year for peak production, but Total executives opted for 2020; in June 2006, CEO Thierry Desmarest advised that, "The capacity of raising (oil) production is a real challenge ... if we stay with this type of production growth our impression is that peak production could be reached around 2020" (Bergin 2006). Apparently, Total advised European states to promote reduced consumption to help ease the situation. Finally, asked whether or not peak oil had arrived in October 2007, the former chief executive of Talisman, a Canadian-based multinational, replied, "We're there or close to it. Mexico, the North Sea and possibly Ghawar are all in decline. The truth is the world is producing 30 billion plus barrels of oil a year and is finding less than 10 billion. This is the worry" (James W. Buckee, quoted in Reguly 2007).

This overview of global statistical data and major company statements highlights Canada's unusual position as a producer that continues to grow—growth that is based on the oil sands—while output of light crude declines. Canada may experience some short-term economic gains as a result of the global situation, but certain facts remain. As a country, Canada cannot escape the pressing reality that oil is becoming both increasingly scarce and increasingly expensive.

Chapter 2

Energy Security

Ensuring a secure supply of energy at a price residents are able to pay is both a difficult task as well as an absolutely critical one. A secure energy supply is vital for a country like Canada, with its relatively small population stretched across a huge area, in a northern landscape that is subject to severe winters.

In the last 50 years, federal energy policy in Canada has oscillated between market liberalism and regulation. This chapter examines the emergence of energy controls as well as the National Energy Policy, developed in response to insecure supply and rising prices. It also looks at the end of the National Energy Policy in the era of deregulation and free trade. The final section considers how security concerns have disappeared from national politics at precisely the time when peak oil is imminent. (This chapter focuses on two of the main energy sources in Canada, oil and gas.)

Minimal Regulation (1950–73)

Until the surge in the price of oil following the Arab-Israeli war in 1973–74, the role of government in Canada was simply to encourage private industry to exploit the country's resources. Canada's supply was cheap, abundant, and apparently unlimited, so energy security with respect to oil and gas did not seem a pressing issue. Moreover, Canada's government was content to allow the country to be developed by private, foreign (primarily US) capital. The oil industry became a prime example.

Canada's oil industry began at Petrolia, southwest Ontario, in the late nineteenth century. Imperial Oil soon dominated the local industry. In 1899, the rising giant from the US, Standard Oil, bought up Imperial. As a subsidiary of Standard and later Exxon, Imperial was the largest company in the Canadian industry through the twentieth century. Although oil was extracted in southern Alberta as early as 1914, production in Canada remained small until the famous discovery at Leduc, Alberta, in 1947. This event propelled major investments, and oil began to flow from the western sedimentary basin. Partly because large-scale Canadian capitalists were uninterested in the risks of oil investment, the new industry quickly fell under the control of multinational companies with non-resident ownership. By 1960, foreign ownership had reached 77.3 percent (Laxer 1983).

Although in the 1960s the Canadian government played a minimal role in the energy sector, one move with long-term significance was the 1961 decision to divide the country into two areas according to their source of crude oil. Alberta's oil fields would supply Canada west of the Ottawa valley, while Quebec and the eastern provinces depended on more expensive imports from Venezuela and the Middle East. This plan worked well alongside evolving US policy: pipelines moved oil south, rather than east to other parts of Canada. This was taking place at a time when the US—recognizing that it was becoming dependent on external oil—was moving to limit imports except from countries with preferred status, including Canada (Laxer 1983). By 1973, roughly half of Canada's production was exported to the US.

The National Energy Board was founded in 1959 with a broad mandate to oversee the development of the industry and protect the public interest. In the early years it usually accommodated the strategies of business without serious disputes. The US majors (that is, the largest oil companies, such as Exxon and Chevron) enjoyed easy access to Canadian oil and exported to the US according to demand. This suited the Canadian government's general orientation to growth based on resource exports. Private enterprise produced and marketed oil and gas. (By contrast, around this time it was primarily provincial governments that generated and distributed electricity. Similarly, nuclear energy—whose history in Canada dates from World War II—has always been produced in government-owned facilities and regulated by the Canadian Nuclear Safety Commission.)

National Energy Board of Canada

The National Energy Board was created in 1959 following a recommendation from the 1957 Royal Commission on Energy. As stated on its website, the Board acts as a regulator of important aspects of the industry:

> the construction and operation of interprovincial and international pipelines;
> pipeline traffic, tolls, and tariffs;
> the construction and operation of international and designated interprovincial power lines;
> the export and import of natural gas;
> the export of oil and electricity; and
> frontier oil and gas activities.

Other responsibilities include:

> providing energy advice to the Minister of Natural Resources in areas where the Board has expertise derived from its regulatory functions;
> carrying out studies and preparing reports when requested by the Minister;
> conducting studies into specific energy matters;
> holding public inquiries when appropriate; and
> monitoring current and future supplies of Canada's major energy commodities.

To meet this substantial mandate, the Board is made up of nine regular members and up to six temporary members selected by government from both the public and private sectors. Board members cannot be involved in the energy business during their terms of office. The Board conducts its work with the aid of over 300 administrative and professional staff. Acting as a quasi-judicial organization, the Board receives and adjudicates applications within its regulatory mandate. Subsequently, the Board oversees projects through their lifespan. It may also offer advice to the federal government and provide information to the public.

Controlled Energy Development (1974–85)

Much greater federal involvement followed the Middle East crisis of 1973–74, when the security of supply to eastern Canada became a serious political issue. Up to this point, the government had no effective control over this core economic resource; following the crisis, there was an attempt to put in place a more interventionist and nationalist policy. This period saw major innovations designed to ensure that the first priority of Canadian resources would be to meet Canadian needs. Although natural resources fall under provincial jurisdiction, interprovincial and international trade are federal responsibilities. This gives the federal government considerable leverage.

Prior to the Arab-Israeli war, OPEC and the majors were pushing up the price of oil. With the outbreak of the war and US support for Israel, OPEC was able to increase prices and declare an embargo on supplies to the US and supporters. Prices jumped dramatically. Saudi Arabian light crude averaged $12.35 in 1974, compared with $6.74 for US crude (Laxer 1983). Suddenly, Canada as well as the US faced the twin problems of high prices and a potentially insecure supply, given that the eastern part of the country had little choice but to import its oil. The Canadian government responded to this situation. In 1973 it moved to cap the price of domestic crude at $4 per barrel and to tax away the difference between this price and the export price to the US. Not surprisingly, the US and the oil companies opposed this strategy. It was advantageous to Canadian consumers and the federal treasury, but it also limited corporate profits: domestic oil was by then much cheaper than imported oil. Also bitterly opposed were the Canadian producing provinces, Alberta and Saskatchewan, whose governments claimed that the federal state was stealing the province's wealth by legislating a selling price that was less than its value in the international market place. Suddenly the oil companies moved from claiming almost limitless resources in Canada to a new position: now they claimed that Canada would not be able to meet its needs from its own resources and would likely become a net importer without major investments to find oil in frontier regions and to develop the oil sands (Laxer 1983). The companies argued that only higher prices would save the situation. Relying on corporate data, the National Energy Board and the government became convinced that there was no choice but to allow the price of Canadian crude to be determined by the international market.

In response to this rapidly changing situation (as well as to nationalist concern about the foreign ownership of natural resources), Canada moved toward direct public investment in the oil and gas industry in 1975. For some time prior to the 1973 crisis, officials and government members had been seriously considering the creation of a national energy company. Such a company would supplement regulation as the conventional means of influence. Regulations are important in this context. They affect a number of key aspects of oil production: how companies behave by setting rules of access to resources; how these resource can be extracted; how labour should be treated; what taxes must be paid; what actions must be taken to protect the environment; and where (and even at what price) a resource can be sold. Regulations make it more or less difficult to make a profit, but it is ultimately a corporate decision whether or not to invest in new assets or to sell an established asset. Investors cannot be forced to invest, and without ownership there is also no way to ensure how profits will be spent.

Dissatisfied with the existing minimal influence of government over the energy industry, in 1974 Pierre Trudeau's minority Liberals—with both determined prodding and the support of the NDP—introduced a plan to create Petro-Canada as a national petroleum company. This national company would permit greater control over oil development (Fossum 1997). The pursuit of profits, rather than the public interest, is what drives private enterprise. Because these objectives may imply different courses of action, the establishment of a state-owned company was a step toward greater control. This move did not, however, threaten the dominating position of private capital in the oil industry; indeed, it was initially introduced as a complement to the activities of established companies. However, when the necessary legislation was passed in 1975, the stated objective was clearly to promote energy security. Recent events had demonstrated that energy should not be left entirely to foreign private companies (Laxer 1983). Petro-Canada was required to act according to government objectives rather than take the pursuit of short-term profit as the top priority.

Soon Petro-Canada was involved in the exploration of new frontier areas in the north (it had received the government's 45 percent ownership of Panarctic[1]) and in the Atlantic offshore areas. In these areas, it behaved much like other private companies, entering joint exploration agreements with Shell and Mobil. In 1976, Petro-Canada started to expand with the acquisition of the Canadian assets of Atlantic Richfield. This was followed two years later by Pacific Petroleum. Petro-Canada became a major vertically integrated producer (in other words, a company that would be involved in all the stages of the oil industry, from extraction to retailing) as well as a leader in exploring high-risk areas (Laxer

1983). After acquiring new processing and retail assets in purchases of Petrofina Canada (1981) and BP Canada (1982), Petro-Canada was highly visible across the country. It was also well received by Canadian consumers.

Faced with the unprecedented spike in oil prices that resulted from the Iranian revolution of 1979, the Liberal government moved to strengthen its influence over oil resources. After defeating Joe Clark's short-lived government in 1980, the Liberals introduced the highly controversial National Energy Program. Oil security was an important objective, with a target for self-sufficiency set for 1990. Canadian ownership of the industry was projected to grow from 30 to 50 percent. A greater share of total revenue would accrue to the federal government (this overall direction clashed with the interests of the producing provinces, as we will see below).

Petro-Canada was encouraged to take over Canadian operations of some foreign companies. Exploration grants were based on percentage of Canadian ownership, so that oil companies could then receive increasingly favourable treatment according to this percentage. Other grants were introduced: consumers were encouraged to convert from oil to alternative heating sources, and companies were encouraged to explore in remote areas. The constitution gave provinces the rights to revenue from resources on provincial lands, but the federal government took advantage of its control of the offshore and northern territories to require a 25 percent share of discoveries in these areas (Bregha 2009). This "back-in" provision was strongly opposed by the industry. By this time, Hibernia had already been identified as a commercially viable field, and the conditions of its development proved contentious.

Business resented price controls, but these pre-dated the National Energy Program as oil export taxes appeared in 1973. Although the price structures proposed in the Program varied for "old" and "new" oil, the overall impact of the new policy was to keep the Canadian (domestic) price below that of the world market. Natural gas prices were also influenced by the specification of minimum export prices. In 1975, the National Energy Board also introduced volume controls. The goal of these controls was to limit exports to amounts that exceeded domestic demand, and these continued with the National Energy Program (Plourde 1991).

Private oil companies, now facing higher taxes and greater controls, opposed the National Energy Program. This protest was joined by the governments of the western producing provinces as well as the US. Many westerners believed that, historically, central Canadian companies and the federal government had exploited the people and resources of the west. No surprise that it was easy to ferment opposition to the National Energy Program by identifying it as a continuation of the extraction of western resource wealth for the benefit of the rest of the country. This sentiment was captured in the notorious bumper stickers: "Let the Eastern bastards freeze in the dark." Nevertheless, for a short time, it appeared that the Program would at least be tolerated after the governments of Alberta and Canada reached an accord in 1981 to allow the Canadian price of old oil (discovered prior to 1980) to rise to nearly $60 per barrel by 1986. Prime Minister Trudeau and Premier Lockheed famously toasted this agreement with champagne (Courchene 2005). This price compared favourably for Alberta with the prior target of $40, but it was also constrained by the agreement that it would not exceed 75 percent of the world level.

Return to the Market (1985–Present)

The 1981 oil pricing agreement between the governments of Canada and Alberta was probably based on a belief that world prices would continue to rise. However, these expectations were shattered almost immediately by subsequent developments: by 1988, oil had dropped to below $20 per barrel. The reason behind this dramatic price drop is related to the complexities of the market. The OPEC cartel could not agree on a production cut to maintain prices in 1982 during a severe recession. A buyer's market emerged quickly when prices tumbled in 1983.

With the agreement between Ottawa and the provinces linked to world prices, the expected increases could not move forward. Moreover, other features of the Program faced ongoing opposition. The Program came under increasing pressure from those whose interests it undermined, like the oil companies. It was also subject to the decline in the price of crude oil after 1982; this seriously cramped the industry in Alberta and reduced pressure on the federal government to control prices. In the end, the National Energy Program lasted for only five years.

Abandoning state ownership and regulation also was influenced by the success of political parties that promoted reliance on the market to counter the growing budget deficits of many states, and the related fear of ever-increasing taxation. Ronald Reagan's administration in the US pressed for the elimination of the National Energy Program: the Program stood in the way of US plans to ensure its own energy supply by enrolling Canada and Mexico as core sources of oil and gas (Rutledge 2005). In Canada, Brian Mulroney's Progressive Conservative Party was elected in 1984. The Conservatives were committed to reducing the role of the state in society, and quickly moved to dismantle the program. This new attitude soon led to the 1985 Western Accord, which allowed all crude oil to be sold at market prices as part of a general trend to deregulation. This arrangement was reaffirmed through the implementation of the 1988 Free Trade Agreement with the US, which prohibited two-price systems. It also restricted the limits that the US could place on imports of various commodities, including oil and gas, from Canada (Doern and Gattinger 2003).

In terms of specifically Canadian security and control of national resources, the Free Trade Agreement placed Canada in a US-dominated security zone. In essence, this meant that Canadian resources became a key contribution to US security. Contracts with US companies could not be rescinded and supply to the US could only be reduced as a result of a declared shortage under specified dire circumstances, such as armed conflict involving the energy. However, the amount of oil or gas shipped to the US could not be less than its proportion of total production in the previous three-year period. The 1993 North American Free Trade Agreement (NAFTA) brought Mexico into the security zone. Mexico, however, was able to reject the proportionality clause (Canada-United States Free Trade Agreement 1988; Doern and Gattinger 2003). In other words, Canada—unlike Mexico—is not allowed to meet its own needs at the expense of shipments that otherwise would go to the US. When the agreement was initially written, Canadian politicians probably believed that domestic oil and gas supplies would last for many decades, and that the main problem was interference with trade.

Only a short time after it had become well established, Petro-Canada became a prime target for privatization in the era of market capitalism that followed in the 1980s. The survival of Petro-Canada became an uphill struggle. Dismantling Canada's national energy policy in favour of open markets like NAFTA included partial, and later full, privatization, of Petro-Canada starting in 1991. Probably fearful that nationalist sentiment might be rekindled, the initial move by the Conservative government placed only 20 percent of the company on the market with a cap on foreign ownership set at 25 percent. The Liberals removed the latter restriction in 2000. By 2005, after the state sold all its remaining shares, Petro-Canada was fully private and functioning as a conventional, Canadian-based, multinational energy company. Four years later, it merged with Suncor under the control of Suncor's management.

Canadian Policy in the Time of Peak Oil

In 2001, President Bush outlined a national energy plan, designed to reduce US dependence on imports from unreliable foreign sources (Rutledge 2005). In this plan, Canadian reserves were treated as part of the US domestic supply. This was a clear implication of NAFTA in operation. And there was a corollary: US security issues became those of Canada as well. Yet energy security has not enjoyed a high public profile in Canada. This is in spite of the fact, as we noted above, that conventional natural gas and oil resources are in decline and that eastern Canada has to import much of its oil.

Figure 2-1 reflects Canadian concern with security following periods of global shortage and high prices. The data on world oil prices—whether in actual dollar values or translated into the value of the dollar in 2008—show that the period from 1973 to the early 1980s put significant pressure on industry as well as on the general public. Prices jumped rapidly after a period of stability, especially at the start of the Iran-Iraq war in 1980. Prices peaked at almost $100 (in 2008 terms). We did not see those kind of prices again until recently, in 2007–8. The first peak in Figure 2-1 corresponds to market intervention and the National Energy Program, while the second peak marks a revival of security and price concerns after 15 years of relatively low prices. The key difference is that supply was restricted in the first spike as a result of politics. On the whole, however, the second spike reflects the arrival of peak oil and increased global consumption

Is the supply of oil and gas a problem in the time of peak oil? Many believe that Canada has adequate internal supplies, given the vast amounts contained in the oil sands. But this source is not only expensive to develop; we will see in Chapter 5 that it is also environmentally damaging to extract and refine oil sands deposits.

Another possibility, much publicized in 2009, is that natural gas from shale can provide a relatively clean transition bridge to

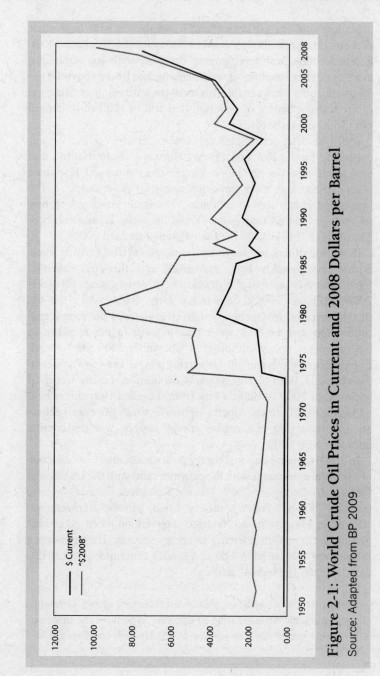

Figure 2-1: World Crude Oil Prices in Current and 2008 Dollars per Barrel

Source: Adapted from BP 2009

other forms of energy; both Canada and the US have large shale deposits (Madslien 2009; Mouawad 2009a; Evans-Pritchard 2009). Recent technological developments do make shale gas viable, but it will take large amounts of water, energy, and prices above $8 per thousand cubic feet to enable this medium-term solution. Shale gas is also a major source of pollution that will be difficult to control even with new technologies.

Some industry commentators are extremely optimistic. For example: Rune Bjornson from Norway's StatoilHydro said exploitable reserves are much greater than supposed just three years ago and may meet global gas needs for generations. ... The breakthrough has been to combine 3-D seismic imaging with new technologies to free "tight gas" by smashing rocks, known as hydro-fracturing or "fracking" in the trade (Evans-Pritchard 2009).

However, others are more cautious, given that existing shale mines have quickly been exhausted and the environmental movement has resisted the development of new ones (Hadekel 2009; Madslien 2009). Canada has large reserves of shale gas in the west and in Quebec. But to date there is no commercial production and no indication that it forms a major plank in federal energy policy (although in November 2009 the National Energy Board did comment favourably, noting the environmental drawbacks). With TransCanada Corporation having received approval in 2009 to build a link from the giant Horn River field in northeast BC to its Alberta network, shale gas may become an important part of Canada's energy supply over the coming decades (NEB 2009).

In 2009, Canada's national energy policy amounted to a collection of historical agreements with the provinces and with the US through NAFTA. In December 2009, Natural Resources Canada's website (entitled "Energy Policy") merely listed various arrangements, referred in vague terms to "targeted intervention when necessary," and made no explicit reference to energy security. The implication is that there is no need in light of Canada's commitment to market-based supply. The website states:

> Markets are the most efficient means of determining supply, demand, prices and trade while ensuring an efficient, competitive and innovative energy system that is responsive to Canada's energy needs (Canada 2009e).

Although reaching the same conclusion, another 2009 document indicates that energy security, broadly understood, does imply "challenges." But this statement classifies Canada as "relatively energy secure" in comparison with other countries, and further notes that Canada's policy is limited to encouraging discussion and "provide an enabling business environment" (Canada 2009b). Canada has resisted any tendency to rate one type of energy as more appropriate and emphasizes the development of the oil sands (Canada 2009b).

This position has been challenged by critics who are embarrassed at Canada's failure to act responsibly against climate change (see Chapter 4) or who would like to see a more independent policy position that places greater priority on Canada's national interests. Laxer (2008, 4) has argued convincingly that "Canada is recklessly unprepared for the next global oil crisis." The Canadian government must establish strategic petroleum reserves and reduce imports to guard against serious interruptions in supply, especially to the eastern parts of the country. This would require the radical steps of redirecting east coast oil for Canadian use and using the Sarnia-to-Montreal pipeline to move western Canadian oil into Ontario— the reverse of what takes place today. It would also challenge the proportionality clause in NAFTA by requiring the same exception that Mexico negotiated (Laxer 2008). It is true that this kind of strategy would involve a significant restriction on market forces. These restrictions would be resisted by corporations, the provinces that benefit from the status quo, the US, and some academics who believe that free markets, technology, and new discoveries are likely to relieve any problems (see, for example, Alvarez et al. 2008).

A Canadian energy policy is difficult to articulate in any area because natural resources are under provincial control; different provinces have quite different material interests according to their endowments. The federal government cannot simply impose a policy on provinces that are strongly opposed, as the failure of the National Energy Program shows. Moreover, NAFTA is a major constraint. So is the US position at any time. Enforcing energy-security measures would likely bring angry objections from all these sources. But at the same time, it is also difficult to see how the government can avoid action for much longer unless Canadians are willing to tie their energy security almost completely to US policy.

Chapter 3

Energy and Equality

Offshore and Arctic resources are more expensive to develop than conventional resources. However, the need for secure supplies of oil and gas, especially following the rise of OPEC and the eventual depletion of inland oil fields in most of North America, have made the economics of these resources look more feasible. But exploiting the natural resources that are located far away from economic and political centres often leads to conflict. Who benefits? And who makes the decisions?

In Canada, both large oil companies and the federal government have been required to negotiate with provincial governments as well as Aboriginal peoples. With a focus on offshore and northern lands, this chapter considers the many players—not just business and government—whose interests have been advanced or undermined in Canada's resource extraction industries.

Offshore Newfoundland and Labrador

The discovery of resources off the coast of Newfoundland and Labrador led to Atlantic offshore development. This is an interesting case that highlights the relationship between resources development and social inequality.

Following its entry into confederation with Canada in 1949, Newfoundland (later changed officially to Newfoundland and Labrador) was perennially Canada's poorest province. Its relatively less-educated population was dependent on resource extraction industries, particularly the seasonal cod fishery. Unemployment stuck persistently at double the Canadian rate in both good times and bad. The discovery of oil offshore opened up the prospect of improving the local economy and bringing Newfoundland society much closer to national standards of material affluence. It was also hoped that it would remove a deeply seated sense of dependency.

Although some drilling took place in the North Atlantic off the coasts of Newfoundland and Nova Scotia in the 1960s, the OPEC embargo and subsequent supply crisis in 1973–74 stimulated exploration. The provincial governments were keen to maximize the benefits from any discovery but faced the problem that the federal government claimed jurisdiction. This became a pressing issue in 1979 when commercial quantities of oil were identified in the Hibernia area (estimates rose from 552 million barrels in 1986 to 1.244 billion in 2008 [Newfoundland and Labrador 2008]) along with natural gas off Nova Scotia. In 1982, Nova Scotia signed a revenue-sharing agreement with the federal government, with the proviso that the province would benefit if any other province negotiated a more favourable deal.

Newfoundland and Labrador however refused to accept these terms and held out for another three years in a struggle to obtain a greater say in the development process. This delay was partly due to provincial and national Supreme Court cases over the legality of Newfoundland's claim to ownership. The decisions went against the province, but eventually a settlement was arranged with the Conservative government of Brian Mulroney (House 1985).

First fields

With the signing of the Atlantic Accord by the two governments in 1985, the Hibernia project could move forward. Newfoundland and Labrador succeeded in negotiating a management system that regulated development through the Canada-Newfoundland Offshore Petroleum Board. (Nova Scotia accepted a similar board structure.) This new Petroleum Board was charged with evaluating proposed projects. In addition to the merit of any proposal, they also considered the following: environmental impacts; local benefits, such as employment; the provision of supplies; and the technology to be used. Oil from the ocean was to be treated for financial purposes as if it were on land. An offset clause ensured that the provinces would receive payments equal to 90 percent of the equalization payments that they would lose as a result of the impact of oil royalties. After five years, this amount would be reduced by 10 percent in each following year (Canada 1985).

But in spite of this agreement, there were several factors that delayed the flow of oil. Oil prices had fallen by the late 1980s, and so expensive offshore development looked less attractive. Moreover, the province preferred a more expensive gravity-based fixed platform, because it could be built locally and would provide much needed employment. It took five more years to reach a practical agreement in which the province reduced its royalties and the federal government committed to pay $1 billion as a grant and to guarantee a further $1.7 billion out of a total investment of $5.2 billion. Gulf Oil, with a 25 percent stake, withdrew in 1992 because of its weak financial position and was replaced by the federal government and Murphy Oil. Finally, in 1997, the first oil was pumped, but under a royalty regime that would produce little revenue. Anxious to obtain employment concessions, the provincial government agreed to royalties that would start at 1 percent and rise only to a maximum of 5 percent by 2004.

Unequal Negotiation: Small Boat Fishers and Oil Companies

"If Hibernia left, I'd feel relieved. ... If it's not there, you don't expect nothing. ... There are so many people depressed and tormented about this that it's sickening" (quoted in Ottenheimer 1993, 198–89).

Several small fishing communities in the early 1990s found oil development closer to a curse than a blessing. The agreement for development of the Hibernia oilfield called for construction of a giant platform. This work had to be next to deep water to permit the platform to be towed to its site. The chosen location, Bull Arm on Trinity Bay, was used by local inshore fishers, whose long-term interests included protection of their fisheries and way of life. How did they fare in their relationship with HMDC, the Hibernia Management Development Company led by Mobil Oil?

HMDC had the financial resources and skilled personnel to wield effective power over fishers of the area who hoped to preserve their way of life. They lost access to traditional grounds with limited compensation, but they did not voice opposition as HMDC shaped their sense of what they wanted. Fishers signed a contract, but under conditions that distorted their understanding of their own interests. The forum for community-level discussion set up by HMDC encouraged questions about opportunities for local employment and business. But matters concerning the fishery and environmental impacts were kept off the agenda. A consultant hired by HMDC to negotiate with fishers did not encourage them to take the formal route of raising issues with the Bull Arm Area Coordinating Committee. Few knew the committee existed. Women plant workers had no voice at all and this was encouraged by the patriarchal domestic structure. Not even their fisher spouses considered speaking for them.

Within a year, fishers learned that their contracts were shaky because the final wording had been changed without them fully understanding the implication that payments would end if the construction project were interrupted. When an interruption did take place in 1992 (after Gulf Oil withdrew from the project), fishers were shocked to discover that they had lost access to their fishing grounds along with the compensation. One fisher put it clearly: "If we had our time back, we would have taken it (the contract) to a lawyer to look over … but we truly thought that Mobil was acting in good faith … now we know it was all a 'put on' to get us to sign." They had allowed others to define their interests and rights. By 1992, fishers were much more cynical: "They don't care about the fishermen; they care about lining their pockets, that's all" (Ottenheimer 1993:156–59).

After signing, most fishers thought they had a good deal with a guaranteed income, although they were told not to discuss the terms with anyone. HMDC's sole right to communicate about the contract was actually written into it. "When we signed that deal we believed them and what they were saying … I really trusted them." When conflict later emerged, this secrecy allowed others in the community to blame the fishers themselves. However, there was much pressure on the fishers: "You know we're not supposed to be talking to you about what's going on … I don't mind telling you about what's going on, but I don't want my name being used in any of this. Lord only knows what would happen if they found out" (quoted in Ottenheimer 1993:155).

The message from this experience is that benefits can be illusory. People less familiar with law and organizational processes need trustworthy advisors to assist in achieving accommodation with oil development. Otherwise, small communities can become even more divided and their long-term sustainability is threatened.[1]

Within a similar framework, other exploration led to the discovery of several new fields. However, the provincial government left the method of extraction up to the oil companies (given that it was no longer willing to subsidize the development). By 1998, 14 years after its initial discovery and following three years of planning and negotiation, Petro-Canada secured an agreement to construct a floating platform for use on the Terra Nova oil field, close to the Hibernia site (Newfoundland and Labrador 2008). This decision meant considerably less local construction and employment. The compensation for the province was that an improved royalty structure would likely double the income to the treasury in comparison with the Hibernia arrangement. Another main field, White Rose, moved forward with Husky Oil as the operator. About one-third of construction expenditures would take place within the province, including more engineering, fabrication, assembly, and integration work than on earlier projects (MacDonald 2001; Newfoundland and Labrador 2003). November 2005 saw the first barrels drawn from the White Rose field. The terms of the Terra Nova and White Rose developments were more advantageous for the province than the Hibernia agreement, but the financial situation of the province has improved very little, and people who live far from St. John's have experienced almost no improvement in their lives.

The Hebron dispute

Conflict and disagreement between the province and the oil companies surrounded the terms of development of the first three offshore oilfields. The oil companies emerged with the best hand, although they were forced to make concessions on technology, employment, or royalties according to the individual case. Discovered back in 1981, Hebron (and Ben Nevis) has still not been developed. Its heavy oil will be more expensive to extract than light crude and the final product will be of lower value. However, with the high prices in 2007–8, and with new fields becoming ever more difficult to find, this location has recently become much more attractive.

As Hibernia and the other fields were being negotiated, little was said about Hebron. By 2000, Chevron, the lead company in the Hebron group, had undertaken sufficient exploratory drilling and evaluation of costs to announce that the field was commercially viable. However, two years of further study led to shelving plans in view of the high costs of development—somewhere between $3 billion and $3.5 billion—during a period of declining prices (Baird 2002). By 2005, with higher oil prices and growing concerns about supply, Hebron again appeared more attractive. But there was a new provincial government in Newfoundland. Danny Williams and the Conservative Party began negotiations. Williams is more a Newfoundland nationalist and populist than a conventional conservative politician. This statement captures his style:

> The biggest thing that we had to change when we came into government was the psyche of Newfoundlanders and Labradorians. We wanted to make sure that they felt very positive about themselves and had self-confidence, that we're as good as the rest of Canadians. ... It's about earning respect and I think we're getting that (Gray 2008).

In this pursuit, Williams refused to be overawed by corporations or other governments. It looked as if an agreement could be reached. However, Chevron announced that it was abandoning the project because the province would not agree on reasonable terms (Chevron 2006). What appeared to rankle the companies was the province's insistence on an equity stake—although the final request was only for 4.9 percent. The second key demand was the introduction of a "super royalty" when oil prices were exceptionally high.

Williams' response to this accusation pointed elsewhere. He claimed that the real stumbling block was Chevron's request for up to $500 million in tax concessions. In fact he identified ExxonMobil as the difficult partner and offered to buy out its share in order to move forward (Jackson 2006). If the company refused to sell, Williams proposed legislation that would limit the amount of time a lease could be held without development. For this Williams would have needed backing by the federal government, but the government refused. Williams was furious: "The fact that the prime minister is not supporting me on the whole fallow field exercise and legislation, the only explanation I can see is obviously he's a supporter of big oil. ... And if he wants to be a big buddy to big oil, that's for him to decide" (CBC News 2006).

In 2007, the situation looked bleak. There was also some public concern that the government had erred by pushing Chevron to the limit and causing a major slowdown in the area's economy. On a number of occasions, the Newfoundland Ocean Industries Association claimed that they would suffer economic contraction with Hebron on the backburner. They urged the Premier to resume negotiations. Williams felt that such comments undermined the government's position by making the "oil companies think they've got wind up their sails ... and the government are going to fold. Well, nothing could be further from the truth" (Baird 2006).

Williams remained firm, oil prices rose, and in the summer of 2007, he announced an agreement in principle that included an investment by Newfoundland and Labrador to secure 4.9 percent of the equity (CTV News 2007a). What had been insurmountable obstacles, according to Chevron, melted away when the profit margin looked more promising. The project was expected to produce about $16 billion for the province over a 25-year period. A year later, the final agreement was greeted enthusiastically by business and the public in St. John's, where house prices reportedly jumped $30,000 to $50,000 the day the deal was proclaimed (*The Telegram* 2008).

This arrangement has many advantages for the province. Long-term benefits, however, depend on an astute combination of reducing the huge per capita provincial debt and investment in long-term, environmentally friendly development. The short-term data carry a mixed message. For example, per capita disposable income shows a strong upward trend in recent years. At the same time, however, Newfoundland has continued to lose population through migration. Moreover, it still has the country's highest per capita debt. The economic benefits of oil are much more visible in the capital city area than in the smaller communities around the coast, which continue to experience economic problems and population decline. Indeed, the St. John's metropolitan area (Northeast Avalon zone) was the only part of the province to gain population from 2001 to 2006; it rose by up 4.5 percent. More remote areas were especially hard hit. For example, the Nordic zone on the tip of the Northern Peninsula, about 1,000 km from St. John's, declined by 12.2 percent (Newfoundland and Labrador n.d.).

Part of the problem for Newfoundland and Labrador in improving the socio-economic position of the province through oil and gas development has been the competing interest of the federal government. The early years of oil production did not suddenly turn Newfoundland and Labrador into a wealthy province with low unemployment. Greater control over all offshore resources was an early objective of the Williams government. His government achieved a much more attractive fiscal arrangement for the province without change in ownership. A weak minority Liberal government—anxious to shore up support wherever possible—agreed in 2005 to a new accord that did not cede ownership of offshore resources to Newfoundland and Labrador (and to Nova Scotia); rather, this accord was likely to provide major new revenue, including an initial payment of $2 billion to be applied against the province's debt (Ommer et al. 2007).

Had it been left intact, the key clause in the new agreement would have permitted the two Atlantic provinces to receive, until 2011–12, offsetting payments to compensate for 100 percent of the amount by which equalization payments would decline. These offsetting payments would continue for as long as the provinces, in the absence of offshore royalties, could be defined as "have nots" entitled to receive equalization payments. If a province still qualified for equalization in 2011–12, and its per capita charges for servicing its debt were higher than at least four other provinces, then the accord would continue for eight more years (CTV News 2007b).

The Conservative Party fought the next national election campaign on a platform that included removing income from non-renewable resources in calculating equalization payments. However, after the Conservative party formed a minority government, the actual legislation that was finally enacted departed from that proposal. Instead, the federal government offered a choice. Equalization payments to a province could be based either on the inclusion of 50 percent of its resources revenue as part of income, or on the exclusion of all resource revenue. However, the rules placed a cap on the amount of money that could be received. A province could choose to follow the existing Atlantic Accord with no cap, but Prime Minister Harper also made it clear that his government would not be extending the Accord after 2012 (CBC News 2007a).

Sentiment against Ottawa was strong: in Newfoundland and Labrador, Canadian flags were lowered on provincial buildings. There was also strongly voiced opposition in other producing provinces. The federal government held firm. Further changes in the 2009 budget reduced significantly the amount that could be received in offset payments by $1.5 billion over the period 2009–12 (CTV News 2009). On the positive side, in 2008–9 the province finally reached a point where it no longer qualified for equalization payments as a result of its oil revenues. In general, the people of the province appeared well satisfied and Mr. Williams' performance as premier was approved by a remarkable 80 percent in February 2010 (Angus Reid 2010).

The North

Aboriginal people form a majority of northern residents. Their interests are critical to an evaluation of resource extraction. Much of this story is about resisting development until the conditions of development are more in line with the perceived needs of local peoples. Whether or not the state has been in a free market or regulation phase, the organization and activities of Aboriginal peoples, aided by environmentalists, have limited the freedom of private capital to develop the industry. This push for an effective voice has taken place in conjunction with attempts to exploit northern resources, particularly the competing proposals to build gas pipelines down the Mackenzie Valley and along the route of the Alaska Highway. Paternalistic colonial rule—without real representation for local people—eventually gave way to more democratic, if slow moving, co-operative structures.

In 1974, opposition to a pipeline from the Mackenzie Delta to northern Alberta led to the federal Berger Inquiry. In the hearings held by this Inquiry, radically different views were presented. Northern business people looked to economic opportunities, while whole communities of Aboriginal people feared the end of their way of life. Reporting in 1977, the Inquiry's core recommendations were that no pipeline should be built for at least ten years and not before the settlement of land claims. These conclusions reflected Berger's view that any pipeline should benefit northern residents, including protection for the cultures of northern indigenous people. Moreover, in no circumstances were critically important habitat areas to be disturbed (Bankes and Wenig 2005; Berger 1988). More than 30 years later, the pipeline remains a source of contention.

There have been recent political changes in these northern territories of the Yukon, the Northwest Territories (NWT), and Nunavut (created in 1999 out of the eastern part of the NWT). Some important land claims have been settled. Also governance has devolved, in varying degrees. With these changes, the responsibility for development is now more complex. The federal state still plays an important role in several incarnations. The federal Department of Indian Affairs and Northern Development retains control over oil and gas management for the NWT, but the Yukon has exercised control of land-based resources since 1998. For its part, Nunavut is attempting to wrest control over development of its "inshore" water resources from the federal government. Nunavut alone is estimated to contain about 11 percent of Canada's oil reserves and 20 percent of its natural gas (Nunavut 2009). However, the main area of interest is in the sedimentary basin of the Mackenzie Delta and western Beaufort Sea (Voutier et al. 2008). All these resources are expensive to develop and lie in an inhospitable, fragile environment.

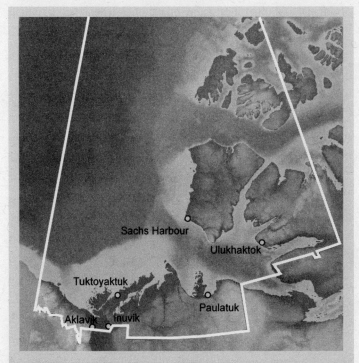

Figure 3-1: The Beaufort Sea Large Ocean Management Region
Source: Canada 2009a

Exploration promoted by the interventionist strategy of the federal government in the 1970s and early 1980s was responsible for the discovery of most of these northern resources. However, with the end of the incentives in the National Energy Program and lower prices in the later 1980s, almost all drilling stopped until the early years of this century. To develop these resources for use involves not only work on the fields, but also the construction of a 1,200-km-long pipeline to reach existing infrastructure in northern Alberta (Voutier et al. 2008).

As of 2010, all northern development goes through stages of public consultation, environmental review, and social assessment. This is required by law. It is no longer possible to simply ignore

Aboriginal and other local interests, although federal and corporate powers do still have sway. For example, in 2004, the Inuvialuit Game Council made the important point that approvals for individual projects may mask the environmental and social impacts of the total development process in the region. So the Council recommended that a strategic regional plan be produced for the Beaufort Sea, which the federal government set in motion with participation from communities, industry, and various levels of government (Voutier et al. 2008). This proved to be a slow process, complicated by the need to co-ordinate with the development of an integrated large ocean management area for the Beaufort Sea, an initiative of the federal Department of Fisheries and Oceans (Figure 3-1).

In April 2008, the report appeared, advocating for local interests that include various forms of collaboration among interested parties, incorporation of traditional knowledge into the planning processes, and careful monitoring of the environmental and socio-cultural impacts of development according to an agreed set of indicators (BSSR Committee 2008). This report formed part of the input into the integrated management process for the Beaufort Sea, the start of which was announced in 2009:

> The goal for the Beaufort Sea integrated management planning process is to have an effective, collaborative process that provides integrated and adaptive management plans, strategies and actions for ecosystem, social, economic, and institutional sustainability. A collaborative process is an open, inclusive and transparent planning, advisory and decision-making process involving all interested and affected parties (BSP 2009).

This inclusive approach is inevitably a slow process, but it does offer some hope that the interests of the people who live in the six communities bordering the Sea will be considered, if not necessarily fully protected. Cases such as the fishers of Bull Arm, Newfoundland and Labrador, show that power differences can still result in manipulation of local understanding that leads to decisions not fully in the interests of weaker participants.

It would be wrong to assert that all northern residents oppose developing their natural resources. With sufficient protection for the environment and a land claims agreement in place, the economic stimulus and royalties from oil and gas development look attractive to many. This includes some earlier opponents such as Steve Kakfwi, premier of the NWT in 2002 (Abele 2005). Some people view this development as an opportunity to escape from dependency and economic colonialism. These advocates include Floyd Roland, the NWT premier, who in 2009 linked the Mackenzie Valley pipeline with community sustainability: "We need to build an economy to build sustainable communities. ... You can have the rest of Canada helping us through transfer payments, or you can have the rest of Canada be partners with us in developing a stronger economy" (quoted in McCarthy 2009a).

Although some local people—especially those who hope for more business—favour development, opposition continues. The pipeline that would enable development has still not been approved. The revival of interest in a new Mackenzie Valley proposal (along with other alternatives) in 2004 resulted in extensive long-term consultation (CBC News 2007b). The consultation process ended in December 2007 after years of work by a Joint Review Panel. At this time, no decision has been reached on whether or not the project will be allowed to go forward. Indeed, the Joint Review Panel did not report to the National Energy Board until December 2009 (Jones 2009). The proposal involved Exxon, Shell, ConocoPhillips, and the Aboriginal Pipeline Group (backed by TransCanada Pipelines) that represented the Inuvialuit, Sahtu, and Gwich'in peoples. This huge project, which would bring arctic gas to the south, may never move forward. This would remain true even if government approval were to come in 2010, following the Panel's report. Some northern business people remain skeptical; there are those who have lost so much that it is now too late. Others, however, are exuberant about their prospects (VanderKlippe 2010).

Frustrated companies have stopped drilling until the pipeline proposal has been approved. The president of MGM Energy, the last active driller in the Mackenzie Delta, condemned the delay: "This is an embarrassment to the country—this project, the regulatory system ... and yet nothing is happening" (Cattaneo 2009). This company made the largest single gas discovery and worked in partnership with several majors. Corporate officials still proclaim their hope for the pipeline. Yet some commentators believe that the partners will not invest the necessary $16.2 billion in this project as long as the price of natural gas remains low and the government does not step in with financial support. This is particularly the case in light of the recent revival of a plan for a competing Alaska pipeline, which might be operational in 2018 (Jones 2009). For example, a Calgary company manager complained, "For this thing to go, the government is going to have to step forward with great wads of cash—many billions of dollars" (McCarthy 2009a).

Context is Vital to Success

In the Hebron dispute, no participant held total power. No participant walked away with everything under request. The Chevron group's key resource was its willingness to develop Hebron. The government's chief bargaining chip was its authority to grant approval of the development plan for expansion into Hibernia South. This appears to have been delayed pending a favourable outcome on Hebron (Cattaneo 2007). Also, Danny Williams' general popularity as a defender of the people remained high. Eventually, an agreement was reached that came reasonably close to the province's objectives.

The provincial government's strategy of holding firm for its preferred conditions was effective, but only because circumstances beyond its control favoured this outcome. In the end, the companies made the decision to return based on their own bottom line:

> [Oil company directors] are always motivated by their ability to make a profit on a project and I think they believe that they can make a good profit from Hebron while still meeting the demands that the province has. So it's basically, sort of, a regular business decision for them.[2]

Corporations do what their managers believe is necessary to make profit. They will eventually reach an agreement when the failure to do so would harm their interests. This appears to have been the situation with Hebron. Had the world been awash with potential oil fields in politically favourable locations, Hebron might have remained on the backburner for decades. Moreover, Canada is considered stable, friendly to capitalism, and thus ideal for investment. The equity stake and increased royalties were moderate when compared with many developments around the world. Mr. Williams' populist style might have irritated some company executives, but was surely irrelevant to the decision to revive the Hebron project.

The proposers of the Mackenzie Valley pipeline have faced a less receptive socio-economic situation since the rejection of the first pipeline and the end of federal exploration incentives. At the time when natural gas prices were high, the second proposal was mired in a slow review process. This appears to have worked in the interests of those local people, mostly Aboriginal, who feel that the development of arctic resources is too great a threat to their way of life. But at the same time, it was bitterly disappointing to those who had invested early in the development process in the anticipation that it would be approved.

Corporation managers, in following their judgement about what best suits the financial interests of their companies, may not proceed with the investment, even if final approval comes quickly. This is simply because gas prices in 2010 do not justify it. Nor does it appear that the federal government will contribute the amount that is required to establish such a vast infrastructure. And competition from the Alaskan proposal remains a factor. Clearly a range of issues, geophysical and social, affect the development in the Mackenzie Delta and Beaufort Sea. The future is uncertain.

By 2010, it is reasonable to conclude that oil and gas development has reduced economic disparity and perhaps social inequality between Newfoundland and Labrador and the rest of Canada. But the benefits are disproportionately concentrated in the capital region. It is uncertain whether or not the government has the will and the capacity to correct this situation. Northern development is less advanced and may stall indefinitely. But should it continue, government policy must distribute the benefits with sensitivity. It must not become a curse for the Aboriginal residents of the north.

Chapter 4

Climate Change

Science has recently linked the burning of fossil fuels since the industrial revolution to climate change. Therefore, the issue of climate change is an energy issue as well as an environmental one. This chapter reviews the science of climate change and the political responses to it, both internationally and in Canada.

What is Global Warming?

Deregulation and free trade marked a strong reduction of state intervention in the economy in the 1980s and 1990s. Working in the opposite direction was a global scale initiative to address the impacts of industrialization on the natural environment. Burning fossil fuels is now seen as the central contributor to global warming and the extreme weather events that accompany it—"global weirding," as it has been called.

The earth's atmosphere contains many different gases—water vapour, carbon dioxide, ozone, methane, and nitrous oxide—that prevent energy from the sun bouncing off the earth's surface back into space. In the absence of these gases, our planet would be much colder. But when the concentration of atmospheric gases increases, so does the so-called greenhouse effect (Figure 4-1). Moreover, evidence from past cycles shows that temperature can vary a dramatic five degrees within a few decades (Maslin 2009).

There is clear evidence that the concentration of greenhouse gases in the atmosphere has risen by some 30 percent since the early years of the industrial revolution (Maslin 2009). It is likely

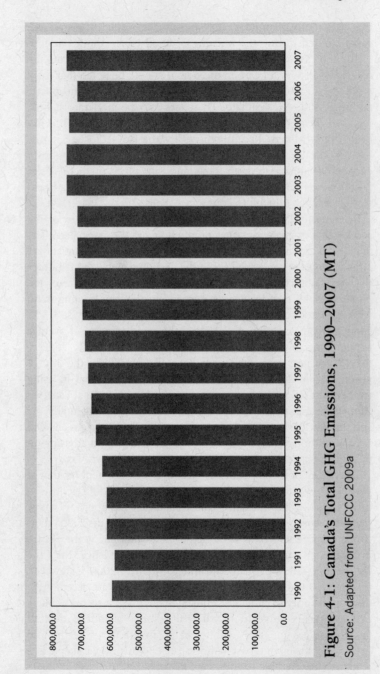

Figure 4-1: Canada's Total GHG Emissions, 1990–2007 (MT)
Source: Adapted from UNFCCC 2009a

that this rapid accumulation is mostly the result of greenhouse gases emitted during the burning of fossil fuels. About one-fifth of the increase results from reduced forest cover, which cuts the natural absorption of carbon dioxide. It should be noted that there is still debate about how much changing land-use and atmospheric pollution will affect climate. The effect may be moderate at first, followed by a "tipping point" once a threshold of resistance has been overwhelmed. Recent reports by the Royal Society in the UK and the US National Research Council point out that abrupt changes in climate are both likely and difficult to predict (Maslin 2009). Since global changes are unlikely to take place on a linear scale, it may be difficult to counteract them.

By the late 1950s, some scientists were already aware of the acceleration of greenhouse gas production and the associated effect of global warming. These scientists were at first less influential than others who were concerned about the declining global temperatures that were observed in the 1960s and 1970s. Some spoke of the possibility of a new ice age. (It turned out that this cooling trend was the product of variation due to the sunspot cycle.) The cooling trend was soon replaced by the rapid increase of average temperatures in the 1980s. Temperatures continue to rise. "Indeed, the 12 warmest years on record have all occurred in the last 13 years" (Maslin 2009). By 1988, concern about the trend in rising temperature led to the formation of the Intergovernmental Panel on Climate Change (IPCC) by the United Nations Environmental Panel and the World Meteorological Organization. The IPCC was charged with monitoring scientific data on climate change and advising on possible impacts on human and environmental systems.

The IPCC does not conduct research, but the reports that it produces—written and reviewed by teams of leading scientists—carry considerable authority. The established credibility of these scientists means that opponents cannot easily discredit their conclusions. There have been some arguments about the IPCC's methodology (including their dismissal of evidence that the medieval period was equally warm [e.g. McIntyre 2008]). However, it is important to note that the IPCC's conclusions are accepted by the vast majority of climate scientists. In any case, it seems unlikely that the recent surge in global temperatures is independent of greenhouse gas emissions. It also seems unwise to ignore this rise over a dispute about past temperatures. While some skepticism remains,[1] few credible scientists are in doubt.

The Kyoto Protocol

Concerns raised by the IPCC were instrumental in the United Nations Framework Convention on Climate Change (UNFCCC) in Rio de Janeiro (1992) and the associated Kyoto Protocol (1997). This Protocol was agreed upon in principal after several years of negotiations. Its objective is to work toward an international agreement to reduce greenhouse-gas emissions (in order to limit the future rise in mean global temperature). Unfortunately, it has proven exceptionally difficult to move forward.

The Kyoto Protocol proposes that a group of 38 industrialized countries—those countries that have historically contributed the most to global warming—would reduce their greenhouse gas emissions over the period of 2008 to 2012 to an average of 5.2 percent below their 1990 levels. There is some variation in specific targets, however. For example, European Union countries have a target set at 8 percent, while the US target is at 7 percent. Other countries with economies in transition have even lower targets. Developing economies do not have targets, but we will see below how they are nevertheless encouraged to invest in clean technology.

Industrialized countries must first and foremost take domestic action against climate change. One proposal is the cap-and-trade system. This alternative involves taking part in market exchanges, where emission credits can be purchased from countries that do not need them to meet their targets. A developing country may have no reduction target, but can nevertheless invest in technology to reduce its emissions. It could then sell these credits to an industrialized country. There is clear financial incentive for all countries to cut emissions. Another part of the strategy is the Clean Development Mechanism, which channels investment into environmentally friendly projects in developing countries in order to earn emission credits. Finally, a country may enter into a joint implementation initiative by investing in another country in order to meet its own target obligation. In some cases, this may prove a cheaper alternative (UNFCCC 2009c).

Although the Protocol was embraced enthusiastically by some countries, others resisted. In order to have the force of law, the Protocol had to be ratified into the national legislative systems of at least 55 of the countries that signed the Convention on Climate Change. Also, the countries signing had to account for at least 55 percent of the total carbon dioxide emissions of the industrialized countries in 1990. President Clinton signed the Protocol, but the US did not immediately ratify it. By 2001, shortly before it became a legal treaty, President Bush repudiated the Protocol on the grounds that it would be too damaging to the US economy. Moreover, he argued that the lack of targets for developing countries was unfair. As the US was responsible for 36.1 percent of 1990 emissions that were coming from the industrialized countries, this was a major blow from which the Protocol could not fully recover. After much persuasion from European countries, Russia's signature in November 2004 put the Protocol across the 55 percent threshold. Having achieved the necessary signatures, the Kyoto Protocol became legally binding on 16 February 2005. By November 2009, 189 countries plus the European Union had signed, but still not the US (UNFCCC 2009b).

Why did the Kyoto Protocol stumble like this? Many countries clearly fear possible downturns to their economies. The dangers posed by climate change have probably also been underestimated. There are many examples of how public opinion has been influenced, allowing governments to back out of support for international efforts like the Kyoto Protocol. An interesting case study of this is the Stern Review. In 2007 top British economist Nicholas Stern produced an extensive report evaluating the cost of climate change mitigation. Stern's study found that 1 percent of GDP invested in 2007 to counter climate change would be minor compared with negative impact on GDP—some 20 percent—if no action is taken before the serious impacts of climate change begin to take hold. But the Stern Review did not influence public opinion as much as was expected. The big energy companies, fearing that they would bear the brunt of change, lobbied effectively against compulsory targets. Until recently these companies helped finance an opposition science that stressed uncertainty. The confusion introduced by those trying to discredit climate change science created sufficient public doubt to permit the US leadership to avoid committing to Kyoto, in spite of the evidence produced by Stern and others. It is interesting that roughly half of the US population has expressed concern about global warming, even if it is not always their main issue (e.g., Dietz et al. 2007; McCright and Dunlap 2003).

Energy Companies on Climate Change

The major oil and gas companies have been strong voices against climate change regulation and effective opponents of government intervention in the US and in Canada (Marsden 2009). Because climate change science pointed to fossil fuels as a critical source of greenhouse gases, major oil companies initially adopted a defensive position, lobbied against mitigating strategies in the key countries, and funded opposition science. After years of arguing and lobbying against the theory that human action was the most important factor in recent global warming, BP, Shell, Total, Chevron, and finally ExxonMobil changed their positions. These companies suggested that they were now taking big steps to address the situation. This change has been awkward for the companies to acknowledge. It has also been paradoxical: even if they accept the science, they are still seeking to expand production, often by drilling in environmentally sensitive areas. And in some respects their current actions are contradictory.

In 2006, BP stated clearly that fossil fuels and global warming are linked. Moreover, this company accepts its responsibility to work toward a solution: "As a major supplier of these fuels it's only right that we play a part in finding and implementing solutions to one of the greatest challenges of this century" (BP 2006). This has been the company's position since 1997. BP was the first major to leave the Global Climate Coalition—a once-influential group that claimed human activity was not the main source of global warming, and campaigned against the Kyoto Protocol. BP's decision to leave the anti-climate change lobby may have helped the others follow in order to share the positive public image.

In 2006, Shell's chief executive appeared strongly committed to dealing with the impact of fuels on the environment: "No discussion of future energy can take place without a focus on the effect of future energy use on the environment. It is an issue of concern to all energy producers, to energy consumers and to governments" (van der Meer 2006, 3). However, he also anticipated that future oil production would generate even more damaging impacts, and thus carbon sequestration and development of cleaner fuels would be vital. Indeed, these strategies are common to all companies. In some ways, Chevron's position is the most encouraging: this company provides public estimates of the final environmental impact of its products (CERES 2004).

ExxonMobil clung longest to the position of denial. After a period of strident resistance to the theory that the global climate is changing—and particularly to the view that burning of fossil fuels is the most important factor in global warming—ExxonMobil's stance became more ambiguous in 2006. Although it acknowledged that certain assumptions do point to serious future consequences of global warming, the company also noted that according to one scenario the warming trend could decline without intervention (ExxonMobil 2006a). The company accepted that fossil fuels were contributing to the process, but demonstrated its faith in technological improvements as the prime means to control greenhouse gas emissions. It also acknowledged that "worldwide carbon emissions are expected to grow rapidly over the next century, even with significant technology advances" (ExxonMobil 2006b, 9). In 2005 the prestigious British Royal Society reported that ExxonMobil contributed $2.9 million to 39 anti-climate-science groups; the Society claimed these groups misrepresented the work of climate change scientists to the general public. In 2006 the Royal Society sent an open letter from member scientists to the company requesting that it stop all funding to groups who opposed mainstream climate science. Finally, in 2008, under pressure from some of its most prominent shareholders, the company acknowledged that its earlier position might have hindered action to counter global warning and that it would cease funding denial groups (Adam 2008).

By 2006 most energy companies had given up the battle against the scientific position on climate change science (Mufson and Eilperin 2006). In the US, with the Obama administration urging action, ExxonMobil advocated a position that once would have been anathema: the introduction of a carbon tax, which chief executive officer Rex Tillerson, in 2008, preferred to the cap-and-trade system. He argued that:

> A carbon tax avoids the costs and complexity of having to build a new market for securities traders or the necessity of adding a new layer of regulators and administrators to police companies and consumers. … A carbon tax is also the most efficient means of reflecting the cost of carbon in all economic decisions … There should be reductions or changes to other taxes — such as income or excise taxes — to offset the impacts of the carbon tax on the economy (ExxonMobil 2008).

However, BP, Shell, ConocoPhilips, and Exxon are also members and financial supporters of the American Petroleum Institute, which led opposition in 2009 to President Obama's climate change bill. Although Exxon was the only one that appeared to favour the Institute's campaign, BP and Shell remained part of the organization (Macalister 2009). With the bill stalled in the Senate, several companies that had been part of an industry-environmental coalition (the United States Climate Action Partnership) withdrew from that Partnership in February 2010 as a result of tension between the positions of the corporate leaders and the environmentalists. Shell was left as the only representative of big oil (Sheppard 2010). The majors appear to be pulled in opposite directions, and their behaviour is sometimes inconsistent.

Co-operation to Address Climate Change

By 2007 it was evident that many countries were far behind their Kyoto commitments. Most remain unlikely to attain them by 2012 (Table 4-1). Only 24 of the 42 Annex 1 countries[2] have achieved any reduction in emissions; the 12 most successful are former socialist states that experienced big declines in economic activity as a result of the dramatic switch to market economies.

Using earlier data, Kerr (2007) analyzed the relationship between emission trends and the adoption of climate change programs. He found that (1) only 7 of 21 countries with such programs experienced declining emissions between 1998 and 2004; and that (2) only four showed improved trends following the adoption of the programs. Although some measures may take longer to have an impact than this time series allows (and there may be other confounding factors as well), Kerr argues that governments such as those in the UK and Germany claim success for policies that is not justified by the evidence. This is a problem insofar as more demanding long-term targets will require much more stringent measures.

In recognition that developing countries are not responsible for most of the existing air pollution and that they should not bear the burden of change, they were not required to reduce emissions in the first stage. The Protocol did however require that an adaptation fund be established to reduce the financial burden on developing countries. Meanwhile, China, India, and several other countries have since become major contributors to global warming as a result of their rapid economic expansion. As Maslin (2009) shows, it will now require a huge effort to limit the actual increase in the next century to no more than 2°C. Most experts anticipate that this objective will fail. Only 18 of 182 experts surveyed by Adam expected to attain the 2°C target (Adam 2009).

Nevertheless, meetings of the committed countries in Bali (2007) and Copenhagen (December 2009) attempted to move forward beyond 2012 with a serious and effective program. It was perhaps expecting too much for 193 world leaders to overcome their differences and develop sufficient trust to generate a new binding agreement in Copenhagen – even with the US now prepared to act under President Obama's leadership.

What emerged instead was a kind of agreement in principle among the developed world and leading developing countries to rein in their

Table 4-1 Greenhouse Gas Emissions in Carbon Dioxide Equivalent, 2007

Countries	2007 GHG	Change from Base Year (1990) to Latest Reported Year (%)
Turkey	372,638	119.1
Spain	442,322	53.5
Portugal	81,841	38.1
Iceland	4,482	31.8
Australia	541,179	30.0
Canada	747,041	26.2
Ireland	69,205	25.0
Greece	131,854	24.9
New Zealand	75,550	22.1
United States	7,107,162	16.8
Austria	87,958	11.3
Norway	55,050	10.8
Finland	78,345	10.6
Japan	1,374,256	8.2
Italy	552,771	7.1
Liechtenstein	243	6.1
Croatia	32,385	3.2
Slovenia	20,722	1.9
Luxembourg	12,914	-1.6
Netherlands	207,504	-2.1
Switzerland	51,265	-2.7
Denmark	68,092	-3.3
European Community	4,051,964	-4.3
France	535,772	-5.3
Belgium	131,301	-8.3
Sweden	65,412	-9.1
Monaco	98	-9.3
United Kingdom	640,273	-17.3
Germany	956,113	-21.3
Czech Republic	150,823	-22.5
Poland	398,881	-30.0
Russian Federation	2,192,818	-33.9
Hungary	75,944	-34.8
Slovakia	46,951	-35.9
Belarus	80,010	-38.0
Bulgaria	75,793	-43.3
Romania	152,290	-44.8
Estonia	22,019	-47.5
Lithuania	24,738	-49.6
Ukraine	436,005	-52.9
Latvia	12,083	-54.7

Source: UNFCCC 2009a

greenhouse-gas emissions and an aim to keep global temperatures from increasing by more than two degrees Celsius, the threshold beyond which scientists say the most devastating impacts of climate change will occur (McCarthy 2009b).

Much remains to be done in 2010, but for the first time countries like the US, China, and Brazil appear to have found enough common ground to move forward. Moreover, the US's own commitment to reduce emissions to 17 percent below 2005 levels will be a useful contribution, if it can be implemented. Canada's apparent resistance to greater action brought harsh criticism from environmental groups (McCarthy 2009b; CBC News 2009a).

Canada and Kyoto

At the same time as Canadian policy was moving away from controlling its own energy resources, the environmental movement was advocating more effective management and gaining substantial public support. Coupled with the massive governmental intervention of 2008–9 to avoid a depression, actions to regulate the production and consumption of energy mark a new period of state intervention—even if pro-market rhetoric is still with us.

Broadly speaking, the environmental movement encompasses all the organizations and individuals who support action to move toward sustainability. From the early 1980s, addressing climate change became a core part of many environmental agendas. Canadians have become familiar with international organizations such as Greenpeace, Friends of the Earth, Earthwatch, and the World Wide Fund for Nature (WWF), as well as Canada-based promoters of a healthy environment such as the Green Party of Canada, the David Suzuki Foundation, the Sierra Club Canada, and the Pembina Institute. Individuals, such as former US presidential candidate Al Gore have also brought the issues to the attention of many. Environmental organizations have formed coalitions, such as the campaign to slow down the development of the oil sands (see Chapter 5). Although it is difficult to quantify how much influence these people wield, many citizens across the globe are expressing ever more concern; these citizens are also part of that movement.

By 2002 Canada moved in an opposite direction to the US by declaring its acceptance of the Kyoto Protocol. Canada's target was to reduce greenhouse gas emissions to 6 percent below their 1990 levels (Brownsey 2005). Similar to positions taken in the US, the ratification of the Kyoto Protocol was strenuously opposed by Alberta and the petroleum industry on the grounds that it would damage economic growth and unfairly excluded industrializing countries. Losses of $40 billion in economic growth and 450,000 jobs were expected by 2010. Buoyed by the failure of the federal government to put forward a specific plan in 2002, a national campaign elicited supporting statements from the premiers of BC and Ontario (Bergman 2002; CBC News 2002). Alberta and the oil companies only supported action to reduce the intensity of emissions; this kind of limited reduction would still allow growth in the absolute volume of pollution as production increases.

Nevertheless, the federal government pushed ahead knowing it had broad support from the public, including Albertans, on this issue. The government may also have feared that a backward move would have serious political consequences. For example, a national poll conducted in May 2002 reported 69 percent in favour of ratification and only 15 percent opposed. Even in Alberta, 54 percent supported the Protocol. (Another larger sample of Albertans from the same period found 70 percent in favour of ratification and most willing to contribute to the costs [Ipsos-Reid 2002].) In addition, more than half of the respondents were in favour of moving ahead rather than continuing debate on the possible economic consequences. Many were also willing to face tax increases and some decline in their standard of living if necessary to support action to limit greenhouse gases (EKOS 2002).

Canada finally ratified the 1998 Kyoto Protocol. If Canada were to take seriously its commitment to reducing greenhouse gas emissions to 6 percent below their 1990 level, then quick and effective action was required. The federal government was not helped by its lack of jurisdiction over natural resources and by the lack of co-operation from Alberta, which remained steadfastly committed to expanding the oil sands. A series of voluntary measures and limited agreements were unlikely to bring Canada close to its target and opened the Liberal administration to charges of mishandling the situation (e.g. Stoett 2008). By January 2006, the situation looked even less hopeful when a minority Conservative government, with at best lukewarm commitment to the Kyoto process, replaced the Liberals.

In these circumstances, what does Canada's record actually look like? Since 1990, the total volume of greenhouse gases from

Canadians Willing to Sacrifice for Environment

CTV News, 26 January 2007

An increasing number of Canadians are willing to make sacrifices for the environment, according to a poll conducted for CTV News and *The Globe and Mail*. About 93 percent of those surveyed said they were willing to make some kind of sacrifice to solve global warming, according to findings from the poll conducted by The Strategic Counsel.

Other notable results include:

> 76 percent are willing to pay to have their houses retro-fitted to become more energy efficient
> 73 percent would reduce the amount they fly to times when it is only absolutely necessary
> 72 percent would pay more for a fuel-efficient car
> 62 percent are willing to have the economy grow at a significantly slower rate
> 61 percent would reduce the amount they drive in half.

About 83 percent of those polled say they feel global warming has the potential to harm future generations. Still, 64 percent of survey respondents said they were not ready to pay significantly higher prices for gasoline or home heating fuel.

Adapted from CTV News 2007c

Canadian sources has actually increased. And it had increased by a lot. By 2007 Canada's emissions—excluding changes due to land-use and forest coverage—were 26.2 percent above the 1990 emissions. In other words, they are now more than 30 percent *higher* than they should have been had Canada honoured its Kyoto commitment. After three years of decline, the sudden rise of 4 percent in 2007 brought the total to the highest on record. Year-to-year comparisons show no clear patterns, except that the trend is upward (Figure 4-1; Figure 4-2). If land-use and forest-cover estimates are included, Canada's performance looks even worse (UNFCCC 2009a).

Energy and global warming are closely connected. Taking all emissions into account, energy production and consumption accounted for 82.2 percent of Canada's total emissions in 2007. This number was slightly higher than in 1990 (UNFCCC 2009a).

Emission intensity—in this instance, the amount of greenhouse gas per unit of production—has declined substantially since 1990. But this progress has been more than offset by the overall growth of population (with the accompanying consumption) and the industries that contribute to those gases (Canada 2008a). Whereas the energy intensity of oil sands production declined from 1990 to 2005, it remained much higher than for conventional oil and natural gas (Canada 2008a). In 2006 processing the oil sands accounted for 4.6 percent of Canada's total emissions of greenhouse gases (Canada 2008a).

A year later Canada's Conservative government, while accepting that this country had performed poorly compared with many others, pointed out several mitigating factors. These include the consumption of Canadian-produced energy in other countries, its northern climate, its population growth, and its economic expansion (Canada 2007). At that time it was clear that Canada would not be able to meet its Kyoto commitments, even by purchasing external carbon credits as well as other available options—options that excluded taking action at the domestic level. The government announced that a domestic solution would be too harmful to the economy:

> The Government's analysis, broadly endorsed by some of Canada's leading economists, indicates that Canadian Gross Domestic Product (GDP) would decline by more than 6.5% relative to current projections in 2008 as a result of strict adherence to the Kyoto Protocol's emission reduction target for Canada (Canada 2007).

Instead, the government hoped to introduce moderate measures that would emphasize efficiency, quickly moving the country toward absolute reductions. Canada announced that it was committed to a reduction of its 2006 levels by 20 percent in 2020, and from 60 to 70 percent by 2050. It is noteworthy that the target for 2020 would still leave total greenhouse gas emissions higher than in 1990, the Kyoto Protocol's base year. The federal government planned to co-ordinate with the provinces, some of which had already established more ambitious plans. British Columbia, for example, aimed to lower total greenhouse gas emissions in 2020 to 10 percent below what they had been in 1990. Ontario's target included a reduction

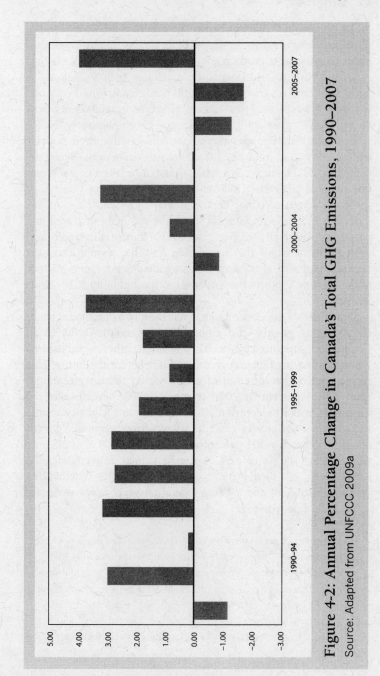

Figure 4-2: Annual Percentage Change in Canada's Total GHG Emissions, 1990–2007
Source: Adapted from UNFCCC 2009a

of 15 percent by 2020. Alberta, the site of massive expansion, remained committed to reduction of emissions' intensity with no absolute cutbacks (Canada 2007).

Canada continues to do poorly in reining in its emissions. In 2009 the Canadian government announced that it would continue to report as required by the Protocol. However, its focus would now be on future agreements, especially the Copenhagen conference (Canada 2009e). Canada's recent position has not been popular among many other countries. In fact, delegates from most developing countries walked out when it was Canada's turn to speak at a recent international conference in Bangkok (part of the preparations for the Copenhagen meeting [Rennie 2009a]). Canada's position was that parts of the Kyoto Protocol should be retained but placed in a new treaty. Although the US and some other industrialized countries shared Canada's view that developing countries should also have binding targets, it is no surprise that Canada's poor performance under the first Protocol has undermined its credibility (CBC News 2009a).

Shortly before the Copenhagen conference, two of Canada's major environmental groups—the Pembina Institute and the David Suzuki Foundation, supported by the TD Bank among others—published a report challenging Canada's record and its plan for the future. This report argued that determined action would reduce greenhouse gases by 25 percent in 2020 (compared 1990 levels), allowing Canada to make its contribution toward limiting global warming to less than 2°C (Bramley et al. 2009). Moreover, economic growth could be as much as 2.1 percent annually from 2010 to 2020, only slightly below the estimate of 2.4 percent without the recommended actions—the so-called business-as-usual approach. Whether the Pembina-Suzuki or the existing federal target is adopted, strong action would be required.

Meeting either target requires governments to put a significant price on GHG emissions (a "carbon price") broadly across the economy, and to back it up with strong complementary regulations and public investments. In this analysis, to meet the 2°C target, a carbon price starting at $50 per tonne in 2010 needs to rise to $200 per tonne by 2020. To meet the government's target, a carbon price starting at $40 per tonne in 2011 needs to rise to $100 per tonne by 2020 (Bramley et al. 2009, iii).

The more radical proposal would still see expansion of oil sands production and per capita GDP would remain much higher in Alberta than in other provinces.

The Pembina-Suzuki proposal would require purchase of carbon credits from other countries, which would lower the cost of the plan and would produce the same reduction in greenhouse gas emissions as comparable domestic action. The proposals also call for major changes in Canada's energy industry. To reach the 2°C target, they also recommended a group of energy-related strategies that were less controversial. These include:

1. Capture and storage of the carbon dioxide generated by the oil and gas industry and power plants, starting in 2016. These technologies are still under development.
2. A decrease in fugitive emissions—i.e., the gases or vapors that leak unplanned from oil and gas production processes and from landfill sites.
3. A general increase in the efficient use of energy in all aspects of private and business life.
4. A greater reliance on renewable energy with wind power accounting for 18 percent of electricity by 2020. Ontario's new Green Energy Act will go a long way in this direction (see chapter 5).
5. The substitution of electricity for fossil fuels (e.g., for heating buildings).

Source: Bramley et al. 2009, 6

The Pembina-Suzuki report's strategy would cost little more in terms of growth and lost income than the federal plan. Even so, however, it was rejected by federal Environment Minister Jim Prentice and the premiers of Alberta and Saskatchewan, who called it irresponsible, divisive, unnecessary, and requiring a transfer of wealth from west to east (Curry and Walton 2009; Thomson 2009). There was no support for either a national carbon tax or a cap-and-trade system. Prentice insisted that the impacts on western Canada could be avoided by co-operation with the US as it develops its own climate change strategy, and that Canada should move forward on a consensual basis. Probably the key to this reaction was the proposed carbon tax that would be borne by the largest emitters and that would funnel back into the economy partly in reduced income taxes for Canadians as a whole. (Other uses would include financing infrastructure for electricity transmission and major improvements to public transportation.) It is true that a greater share of the cost of change would come from Alberta on a per capita basis, but this may be justified on the grounds that Alberta also contributes more on a per capita basis to the problem.

The Pembina-Suzuki plan was opposed not because it was a poor analysis, but because energy companies want to maximize shareholder returns in the short-run, and the government refuses to address the environmental costs of its development strategies. Canada's government has experienced considerable public pressure on the climate change problem. Prime Minister Stephen Harper decided to go to the Copenhagen meeting following President Obama's decision to attend. Nevertheless, Canada had no independent position to promote, having decided to do no more and no less than the US:

> The prime minister says Canada's goal for reducing greenhouse gases is "virtually identical" to targets proposed by the Obama administration. So he says major changes at this stage would put Canada at a disadvantage with its biggest trading partner (Rennie 2009b).

At present (April 2010), it appears unlikely that the federal government will move aggressively in areas such as carbon taxation. In Canada, however, provincial governments also have a substantial impact on energy and climate change issues. Already, both BC and Quebec have moved forward with carbon tax plans. Alberta taxes emissions of large companies when they exceed 100,000 tons annually, but the province continues to reject any discussion of a federal tax (CBC News 2008).

The vast majority of climate scientists expect that global warming will accelerate in the present century with serious environmental and social consequences unless we take action. There are signs of hope in the actions of some governments, but global co-operation remains elusive.

It is disturbing, to say the least, that the lack of a decisive international agreement leaves us turning our backs on the scientific evidence.

Chapter 5

The Oil Sands

Alberta's oil sands (sometimes called the tar sands) are a focus of some of the major issues about energy in 2010. Figure 5-1 shows the size of Canada's oil sands deposits—comprising an area larger than England. Extracting oil from the mix of bitumen, sand, and clay that comprises the tar sands is not only environmentally destructive on a local scale. It is also a process that generates four times the amount of greenhouse gases as regular oil. Less widely considered are the social and cultural impact on both Aboriginal peoples and settlers in the areas where extraction and processing take place. This chapter provides an overview of the contentious aspects of oil sand development. Do the benefits justify the costs?

Overview

Alberta supports a range of alternative and renewable energies in addition to its natural gas and conventional crude oil industries. The fact remains, however, that the largest component of its energy production—and the one with the most resources for the future—is bitumen extracted from the oil sands. Conventional oil production peaked in the 1970s and natural gas in 2001 (Alberta 2008). As we saw above, new discoveries tend to be small and do not keep up with production. No surprise that Albertan expectations and hopes to remain a major energy producer on a world scale lie with the oil sands. We noted in Chapter 2 that the oil sands are the primary reason that Canada's overall production is growing.

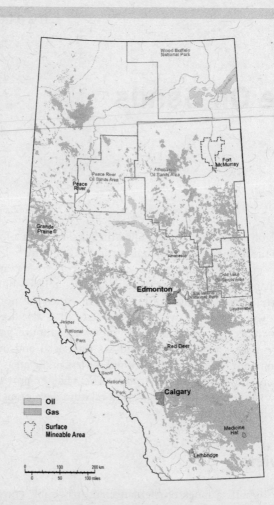

Figure 5-1: Alberta's Oil Sands
Source: Alberta 2008

On a global scale, the size of Canada's proven reserves are second only to those of Saudi Arabia. This is due almost entirely to the oil sands in Alberta (Figure 5-2). In 2004 Alberta estimated that 174 billion barrels (11 percent of the estimated resource) spread

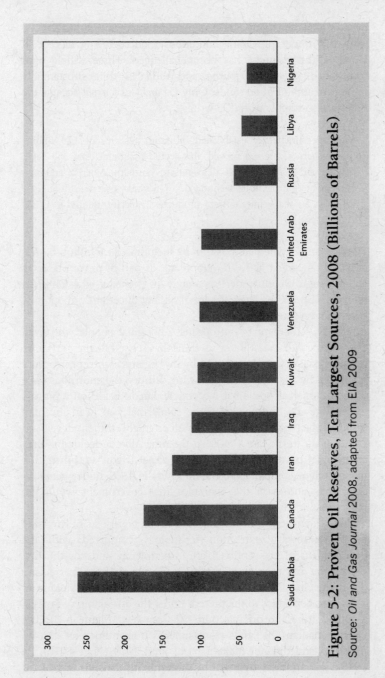

Figure 5-2: Proven Oil Reserves, Ten Largest Sources, 2008 (Billions of Barrels)

Source: *Oil and Gas Journal* 2008, adapted from EIA 2009

over an area of roughly 80,000 square kilometres were likely to be recovered under prevailing conditions (Söderbergh et al. 2007). However this estimate has been challenged. Three widely used sources—BP, *Oil and Gas Journal*, and *World Oil*—differ substantially in their treatment of oil sands. Only *Oil and Gas Journal* accepts the Alberta assessment. *World Oil* writes:

> Alberta's estimates of established oil sands reserves of 174 Bbbl [billion barrels] are not proved; that would require at least 350 Tcf [trillion cubic feet] of gas delivered to northern Alberta, and/or implementation of future technologies. Oil sands reserve estimate is based on 50 years times current production capacity (quoted in EIA 2009, note 4).

BP takes an intermediate position by including 24.5 billion barrels of oil sands "under active development" as part of proven reserves and listing the remainder separately as "remaining established reserves" (BP 2009). There is a lot of oil, but uncertainty about how much can be extracted.

Prior to settlement by Europeans, Aboriginal people who lived close to the Athabasca River used oil that seeped to the surface in certain areas. It was also noticed by the region's eighteenth-century explorers. By the end of the following century, the enormous scale of the bitumen deposits had become apparent. Based on a process to separate the bitumen from the sands that had been invented earlier in the twentieth century, the Albertan government developed a pilot plant in the late 1940s. At the time Alberta decided against commercial production. Instead, the Social Credit administration invited major companies to the oil sands in 1951 under generous conditions and with the assurance that government would no longer be directly involved (Pratt 1976).

No rush ensued, probably because of the high costs of investing in the necessary extraction technology compared with the development of conventional crude petroleum. Indeed, the first production only took place in 1967 at the rate of 45,000 barrels per day in the Great Canadian Oil Sands plant north of Fort McMurray. This plant continues to the present under the ownership of Suncor, a spinoff from the original Sun Oil Company (Figure 5-3). The Great Canadian Oil Sands plant suffered from numerous start-up problems and did not make a profit until 1974 (Pratt 1976). By 1978 a second plant, operated by the Syncrude consortium, started

Figure 5-3: Suncor Tailing Ponds
Source: Pembina Institute

up at Mildred Lake following many years of politics and planning that dated all the way back to 1964. The high costs of producing this oil discouraged further investment until the opening of the Shell Canada plant at Muskeg River in 2003. This event was the first in a major wave of investment. By 2008 all plants averaged 1.31 million barrels per day with growth projected to reach 3 million by 2018. In 2008, 59 percent of the bitumen was upgraded in Alberta, with the remainder exported for processing (Alberta 2009a).

Oil sands consist of varying proportions of bitumen, sand, silt, clay, and various heavy minerals (Figure 5-4). Rather like molasses, the bitumen is a thick, black, tarry material of high density and viscosity that requires extensive processing before it can form the feedstock for refineries. The upgrading process involves adding hydrogen or removing carbon, and depends on natural gas to supply heat and steam. The open-pit mining method of obtaining the bitumen requires stripping away all surface material until the oil-bearing sands are exposed. This may involve up to 75 metres of excavation, after which deposits are moved to a plant where the bitumen is separated, leaving the remainder as waste known as tailings. The process requires large amounts of water.

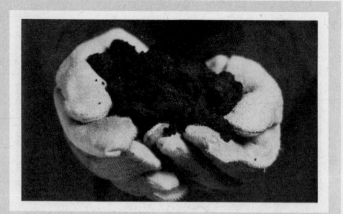

Figure 5-4: Oil Sand
Source: Suncor Energy

Open-pit mining, the method adopted in the earliest plants, is only possible under current conditions for about 20 percent of the resource, much of it situated close to the Athabasca River. The remainder requires an in situ process; in other words, the properties of the bitumen must be "changed" before it can be removed. Its viscosity (or thickness) must be reduced, usually by injection of steam generated from natural gas. This is an energy-intensive process that permits extraction of only about 20 to 25 percent of the bitumen, although more efficient technologies under development are promising (Söderbergh et al. 2007).

Oil sands production is extremely expensive, involving multi-billion dollar projects that often exceed their budgets. Such projects are only viable with high oil prices; during the recession of 2008–9, expansion plans were cut back. In 2007 existing and planned investments amounted to approximately $60 billion (Söderbergh et al. 2007). In 2008 further investment amounted to a record $19.2 billion. Five plants became fully active (Alberta 2009a).

Of course, with North America so hungry for oil, many are delighted with the prospect of mining Alberta's huge deposits. Yet the environmental and social costs are also high. Protest has been growing. Such voices are not new: Larry Pratt's critique, written in 1976, could have been written today.

Since the federal government's withdrawal from the National Energy Program, it has not used its right to review environmental impacts in a way that would challenge Alberta's plans. Alberta's legislation is the key policy framework. The provincial energy strategy (Alberta 2008) does indicate concerns about long-term sources of energy, current environmental issues from both production and consumption of fossil fuels, and even the social impacts. The Minister of Energy states:

> In our vision for our energy future, Alberta will remain a global leader, recognized as a responsible world-class energy supplier, an energy technology champion, a sophisticated energy consumer, and a solid global environmental citizen. ... Clean energy production will be achieved through the application of energy technology leadership such as our government's investment in development and implementation of gasification technology and carbon capture and storage. In a world counting on energy from all sources, Alberta's advantage lies in being able to produce and consume fossil fuels in a far cleaner way, but our commitment extends to the increasing role of alternative and renewable energy (Alberta 2008, 2).

This vision is commendable but rather optimistic: it relies on undetermined technological fixes such as carbon capture and storage. The relevant government agencies have rarely delayed, let alone prohibited, industry proposals. Alberta's strategy also attempts to direct concern about climate change from industrial producers to consumers, stressing that there are more emissions created from oil consumption than production. Of course, Albertans and others can act to improve consumption practices. However, the fact remains that as a relatively small consuming population and a large producer, Alberta's greatest impact would be to exercise control at the level of production. There is no hint in policy statements that the government would consider intervening in the expansion of oil sands production until solutions have been found for some of these problems.

Environmental Impacts

Environmental activists have been warning for years about the impacts of oil sands development. One of the most important environmental issues concerns the volume of water that the oil sands require. Between 2.5 and 4 barrels of water are needed to extract one barrel of oil for raw bitumen (NEB 2008). Although this process is becoming more efficient, the total volume employed in oil production is twice as much as the city of Calgary requires (Woynillowicz and Severson-Baker 2006). Much more has been approved for planned projects. Local people are alarmed.

Most of the water comes from the Athabasca River system (Figure 5-5), which raises concerns about impacts on the river's fish and the people who depend on it for their drinking water. There has been opposition from local Aboriginal people and from environmentalists. In 2009 Greenpeace caught the headlines by occupying three sites in protest at what they claimed to be the crime of dirty oil production. On 30 September 2009, 21 protesters were charged with mischief to property after blockading and stopping two conveyer belts for a day at Suncor's mine (CBC News 2009c).

The Canadian Association of Petroleum Producers (CAPP) defends the environmental record of oil sands producers against the charges laid by environmental critics in several ways. In terms of water, the Association argues that "many projects recycle about 90 per cent of the water used in their operations and use deep-well saline water as an alternative to freshwater where possible. Other projects use techniques that use little or no water" (CAPP 2009, 3). Although the oil sands production process at all plants operating in 2009 takes only about one percent of annual flow from the Athabasca system (CAPP 2009), expansion is a real concern. The oil companies claim that there is no threat to the river or local populations and that they are following government regulations. But many remain concerned. Water use is more of a problem in winter when the plant requires the same amount of water but the river's flow is reduced (Woynillowicz and Severson-Baker 2006). A review by Alberta's government states, "Over the long term the Athabasca River may not have sufficient flows to meet the needs of all the planned mining operations and maintain adequate in stream flows" (Alberta 2006, 112). Clearly water scarcity is an issue and may limit plans for expansion unless more efficient technologies are found to be effective and economically viable. (An example of this is the underground controlled combustion alternative at the Whitesands pilot project.)

Natural gas burns cleaner than oil and is widely used in Alberta and elsewhere for generating electricity, heating, and household

Figure 5-5: Athabasca River
Source: David Dodge, Pembina Institute

appliances. Large amounts of natural gas are used in the oil sands extraction process, especially at in situ sites. The National Energy Board reported in 2008 that:

> It takes about 28 cubic metres (1000 cubic feet) of natural gas to produce one barrel of bitumen from in situ projects and about 14 cubic metres (500 cubic feet) for integrated projects. Currently, the oil sands industry uses about 17 million cubic metres (0.6 billion cubic feet) per day of purchased gas, or about four percent of the Western Canada Sedimentary Basin production. By 2015, this increases to about 40 to 45 million cubic metres (1.4 to 1.6 billion cubic feet) per day, or nearly 10 percent, assuming gas production stay level at 467 million cubic metres (16.5 billion cubic feet) per day (NEB 2008).

This is also optimistic. The production of natural gas may very well not remain stable; in fact, Albertan production has been declining since 2001 and the future of shale gas is uncertain. This suggests higher prices for consumers in the long-term, even without considering the requirements of the oil sands plants. Moreover, environmentalists question the use of a relatively clean fuel to obtain gasoline through a dirty production process. It is possible that the

Alberta government may feel compelled to support the construction of nuclear reactors as an alternative, but this still remains under consideration.

In Chapter 4, it was noted that the single largest source of greenhouse gas emissions in Canada is the processing of oil sands. In 2006 processing the oil sands accounted for 4.6 percent of Canada's total emissions of greenhouse gases. Whereas the energy intensity of oil sands production declined from 1990 to 2005, it remained much higher than for conventional oil and natural gas (Canada 2008a).

Canadian in situ production, which will become more common, generates more greenhouse gases than surface mining. The cleanest in situ processes add roughly twice the volume of greenhouse gases when compared with the most polluting sources of conventional oil. Moreover, the Alberta government's target to reduce the intensity of emissions, coupled with planned expansion in the oil sands, will still mean absolute increases in 2050 compared with 1990, even with a downward trend after peaking about 2020. Unless intervention takes place or some unexpected clean technology becomes available, the oil sands will take a growing share of Canada's total estimated emissions—up an astounding 272 percent by 2020 when they will account for 44 percent of the increase and 12 percent of all emissions (Grant et al. 2009). It will also confirm Alberta as the province with the largest contribution to that total, well ahead of Ontario (Canada 2008a).

Even if most future development would be underground, the oil sands underlie so much of Alberta's boreal forest (about 37 percent) that a great deal of this important natural environment would be harmed. The damage wrought by surface mining is obvious. So is the conversion of over 50 square kilometres of land to ponds that hold toxic tailings waste. Although the industry organization (CAPP 2009) claims that that reclamation is moving ahead successfully, the Pembina Institute, among others, has pointed out that only one square kilometre has been certified as fully restored after more than 40 years of operations (Grant et al. 2009). Moreover, the same report, while agreeing that in situ processes are less destructive of the immediate landscape, argues that "Recent research reveals that the land area influenced by in situ technology is actually comparable to land disturbed by surface mining when fragmentation and upstream natural gas production are considered" (Grant et al. 2009, 21). In other words, the area affected is likely to extend well beyond the boundaries of the production sites. Outside observers find it difficult to verify company claims when there are no clear standards as to what counts as full reclamation and no accessible data on what has actually taken place.

Socio-economic Impacts

The environmental problems connected with the development of the oil sands have been publicized on a national, and even on a global, scale. The socio-economic impacts are less visible. It is rarely clear whose interests the oil sands serve.

The companies that extract and process oil are the most obvious beneficiaries of this development. They have access to a resource that promises profit for decades if oil stays above $70 a barrel. If the companies prosper, that means that their owners and employees can expect continued economic benefits. As they are public corporations, this means that individual shareholders and owners of mutual funds or pension plans that invest in these companies are also beneficiaries. These networks of ownership extend far beyond Canada's borders. Similarly, owners and employees of companies that provide services or goods used in extraction, processing, and distribution will gain. Residents of Alberta may continue to pay less tax than other Canadians as they benefit indirectly from the revenues that accrue to the provincial government. Employees who depend directly or indirectly on oil sands development often migrate from depressed economic areas; many maintain a residence in their original home region (or eventually return there). In either case, there are likely to be benefits to those local economies. In general, consumers of energy benefit even if the ultimate cost to them is high.

With all these beneficiaries, it may seem at first glance that the only real issues are environmental. Yet even the former premier and enthusiastic promoter of Alberta's oil sands, Peter Lougheed, raised concerns. In 2006 he noted the use of valuable—and increasingly scarce—natural gas to extract the bitumen may turn out to be extremely foolish in the long term. But he raised other concerns as well: for one, there is little return to the provincial treasury from new projects until the burgeoning costs of development have been covered. He was also critical of the hospital and schooling situation in Fort McMurray, and concluded:

I keep trying to see who the beneficiaries are. Not the people in Red
Deer, because everything they have got is costing more. It is not the
people of the province, because they are not getting the royalty return
that they should be getting, with $75 oil (Lougheed 2006, 5).

Lougheed warned that the provincial economy was overheated,
that poorer people and new migrants would suffer from high living
costs, and that young people would abandon their education for
highly paid, relatively unskilled jobs connected to oil development.
Other critics have raised similar points. It is worth noting that until
recently, critics like these were seen as "radicals."

Aboriginal people have enjoyed various benefits: some companies,
for example, have actively sought to employ Aboriginal workers.
Recently, for example, Suncor and the Fort McKay First Nation
announced a joint business resources centre (Nation Talk 2009).
According to information supplied by the government of Alberta,
in 2007 more than 1,500 Aboriginal people worked directly in oil
sands production while others were employed in construction.
Moreover, in that year companies owned by Aboriginal people held
contracts worth $606 million (Alberta 2009b).

Despite such gains, by the summer of 2008, chiefs from
three provinces and the Northwest Territories, meeting at Fort
Chipewyan, came together in a plea to halt further development
until an assessment of how to mitigate the negative impacts is
complete (Henton 2008). As Chief Allan Adam put it:

We have to slow down industry to let us catch up. ... If we continue
to let industry and government behave the way they've been behaving
the last 40 years there will be no turnback because it will be the total
destruction of the land (quoted in Henton 2008).

Their primary concerns are impacts on the water quality of the
Athabasca River and the health of the people. The chiefs argued
that development was moving ahead without their consent and
that requests for baseline health studies were being ignored. They
proposed alliances with others groups, along with negotiation
and legal action to protect their interests. Some months later, two
of these First Nations' groups combined with the Forest Ethics
environmental group to place a full page advertisement in USA Today
criticizing the continued development of Canada's "dirty oil" (CBC
2009d). This should be seen as part of a broader campaign by 16

Canadian and US environmental organizations to put a moratorium on development in part by bringing pressure from President Obama and the US public, which consumes much of the oil (CBC 2009e).

By September 2009 the struggle had reached the courts. The Chipewyan First Nation challenged leases granted to Shell, claiming since the First Nation had not been consulted as guaranteed by treaty the leases were invalid. Alberta argued that the opposition was too late, but lawyers for the Chipewyan countered that they had not been informed about the lease, which is part of the development process. This raised the critical issue of when consultation has to take place. Government guidelines argue that this is not required at the leasing stage (Zabjek 2009). The court case continues.

Apart from Aboriginal people, many other residents of northern Alberta—even those who earn high incomes as a result of oil sands development—have reservations about the social costs of rapid development. The article in the text box below, written just before the downturn of 2008, indicates that the economic gains in Fort McMurray are tempered by high costs of housing and services. Physical infrastructure such as roads, schools, and sewage treatment lags behind population growth, which was expected to jump from 65,000 to 100,000 in only three years.

Boomtown Hustles to Keep Up with Breakneck Growth

J.A.C. MacDonald, *Edmonton Journal*, 26 March 2008

With $20 billion slated for oilsands investment this year alone, job-wise, housing-wise and business-wise, there can be no doubt Fort McMurray is on fire.

"It's exciting. There's lots of action. People have a pretty good attitude," says Milly Quark, president of the Fort McMurray Real Estate Board.

However the immensity of development may be lost on many Albertans and Canadians. The $20 billion represents more money than will be invested in manufacturing in the rest of Canada this year and a sum that is almost double the annual rate of the previous few years. It's a staggering amount as companies work hard to extract energy from the world's second-largest oil deposit from lands north and south of this city of 65,000.

Despite the pressures, the city is working hard to become a wonderful place to raise a family, with the $140 million MacDonald Island Redevelopment Project underway, and a new sports complex recently opened by Keyano College.

New suburbs and modern new homes are going up quickly, and land is being readied for new houses of worship, including a Catholic church and a mosque.

The stereotypical partygoer at a downtown tavern is less representative of Fort McMurray than busy families with a growing quality of life, or a single older mobile worker who can be found reading a book over coffee at Tim Hortons, says local businessman Dave Kirschner.

Statistics back up Kirschner's observation. Most of the 25,000 mobile workers living in camps at oilsands construction projects or in the city are men aged 35 or older. More of them are married than the Alberta average, a 2007 study shows. But this is a place with housing and infrastructure pressures. Even for a real estate agent, the price of a single family house is getting to be too much. ... More land is needed yesterday for servicing and development to keep up with a population that is conservatively expected to grow to 100,000 within three years, says Jacob Irving of the Athabasca Regional Issues Working Group, the local voice of the oilsands industry. ...

But people can still gain, because this is the land of opportunity, residents say. Plans and construction abound for building and development to increase quality of life in and around Fort McMurray. They include a new bridge over the Athabasca River, and $100 million for an enlarged airport to help cope with an average 195 flights a day—70,802 flights a year—at Fort McMurray airport last year. ...

And it's all prompted by oilsands plants going 24-7 north and south of town. Some plants are so big they have huge camps of workers on site and their own runways to accommodate large jet aircraft. After much discussion, work camps (the municipality prefers to call them project accommodations) were also permitted inside city limits for workers at major construction projects such as the MacDonald Island redevelopment.

Source: Adapted from MacDonald 2006

Chapter 6

Electricity

Supplying electricity to homes and businesses takes place in stages: production, transmission, and distribution. The operation of some of these stages in Canada has gone through considerable change over the last 50 years. Since 1990, there have been various degrees of privatization and deregulation, mostly in the distribution and production sectors. We have seen three related areas of contention: a secure and safe supply of electricity, consumer price regulation, and intervention to address issues of pollution.

Global Overview

Electricity emerged rapidly in the late nineteenth century, initially dominated in most advanced industrial countries by large manufacturing and financial companies. These companies also extended their networks into many less developed societies. It was not long before electricity formed the core component of any given society's basic infrastructure. Given its importance, there was growing pressure for electricity to be controlled nationally. Multinational business enterprises were replaced by public ownership: "Over time, around the world, countries wanted their power plants to be run by nationals and owned domestically. Thus, gradually and globally, multinational enterprise in this sector disappeared" (Hausman et al. 2008, 274). This process was largely completed by the late 1970s (Beber 2003; Hausman et al. 2008); some years ago, however, this was reversed as part of the neo-liberal stress on the markets and deregulation.

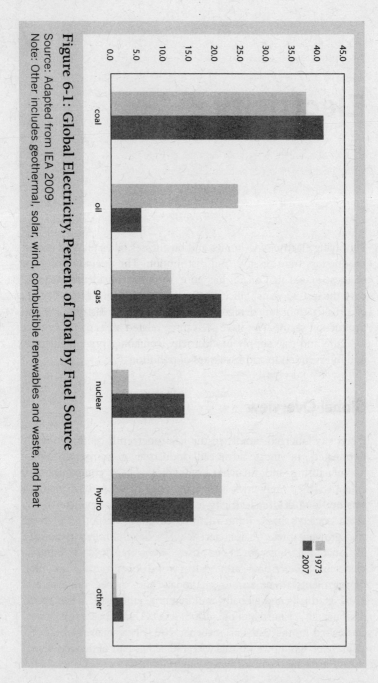

Figure 6-1: Global Electricity, Percent of Total by Fuel Source

Source: Adapted from IEA 2009

Note: Other includes geothermal, solar, wind, combustible renewables and waste, and heat

The consumption of electricity has grown rapidly in recent decades, even as its sources have changed. Between 1973 and 2007, global production rose roughly 300 percent. Over this time, electricity generation became far less dependent on oil than it had been in the 1970s (Figure 6-1), falling from 24.7 percent to 5.6 percent. In the same period, the generation of electricity from natural gas rose from 12.1 to 20.9 percent. Coal also increased slightly, to 41.5 percent. Fossil fuels, then, still account for about two-thirds of electricity production. China's dramatic growth has been fuelled more by coal than any other source of power; China itself produces about 40 percent of the world's coal. The other notable change in recent years has been the use of nuclear energy to produce electricity. By 2007 nuclear energy had reached 13.8 percent of the total produced, dwarfing any increase in energy from renewable sources. Overall, electrical power has remained largely dependent on non-renewable sources and continues to be a major generator of greenhouse gases.

Markets, Supply, and Price

A consistent supply and affordable prices are ongoing concerns of industry, organizations, governments, and residential consumers. Since the 1980s, privatization has been a key word in the supply of electrical energy. The allocation of electrical power through a competitive market has been promoted as the only way to assure reliable, cost-efficient electricity. For some people, ownership is ideological rather than practical. Some believe that energy is a public service that should be supplied at a low cost to all consumers based on joint public ownership. Others believe that the general good is best served when private entrepreneurs compete as the means to generating more efficient production and lower prices.

At first glance, the market model—based on competing private enterprise—is the most attractive to consumers. In this model, consumers would have the option of adjusting the amount purchased according to the price of the electricity. However, electricity is an essential item that consumers must purchase at a minimum level regardless of its cost, and at the expense of other desirable commodities. Conservation helps, but it is also limited by the initial cost of more efficient technology. Electricity markets with few suppliers can easily return to monopoly conditions, and sellers can also impact prices by withholding supply. In both Alberta and Ontario, market experiments floundered when the price of electricity rose rapidly, and politicians felt compelled to respond to public unrest by controlling prices.

Advocates of the market model argue that public (or government) employees are unable to make cost-efficient judgements, or that they have no motivation to do so. It is easy to find examples of poorly managed government business. It is equally easy, however, to find the same in private companies. Businesses do fail, and on a large scale—recall the case of Enron. Many public employees are dedicated to their jobs, have a personal interest in lower rates for services and lower taxes, and are excited about new technologies. And in a similar vein, middle- and lower-ranking employees in the public sector have no more or less motivation to work hard than do those in the private sector. In fact there is evidence that improved job security makes them less anxious, more loyal, and therefore more productive. Fear has not been shown to be an effective motivator.

An intermediate position here is to allow private enterprise, but keep it subject to overall public regulation. This arrangement is seen, for example, in situations where some part of a public monopoly of generation is sold, or where new investors are encouraged to contribute to the power supply. Of course any investors in this kind of an arrangement must be able to anticipate profit; in such circumstances, profit can be assured in the form of long-term contracts between power suppliers and the public distributor. Taxpayers will pay higher prices, but the state—and therefore the public—are relieved of the burden of financing new developments. In Ontario and Alberta, private retailers have been brought into the system by allowing bidding for supply from the generators. However, the limits that were placed on retail prices after the initial price surges are probably discouraging to investors.

What does an electricity market typically look like? Generation, transmission, and distribution are separate components of the power industry. Generation and transmission involve high capital costs for technology, which also makes large-scale production more efficient and discourages competition. They form easy targets for a market domination strategy. For these reasons, electricity generation and transmission are often state-owned or at least tightly regulated. Generation may be partly privatized and private capital may be encouraged to invest in new plants. Local links in the distribution network may act as direct sellers to consumers or as wholesalers for private intermediate sellers. In Ontario both conditions apply as the local distributor supplies all customers who do not make a contract with a private supplier. Market prices are established at hourly or smaller intervals according to supply and demand. This fluctuation results from the inability to store electricity after its generation—it has to be moved and used (Dewees 2005).

There are many people who prefer the deregulation of energy—in other words, those who support private enterprise instead of government ownership. In the private enterprise model, for example, prices are set according to supply and demand. The argument in favour of private enterprise is that energy will be supplied more efficiently and at a cheaper price than a system based on public enterprise. The claim about efficiency is based on the pursuit of profit: the interests of profit will lead business managers to choose the smartest, cheapest, and most productive forms of technology. In this model, lower production costs will translate into lower costs to consumers (as well of course as higher profits). The competition among providers ensures that the best system will prevail. In this model, the market must contain several suppliers who could in principle meet the demand. This model also assumes that no supplier can manipulate the market by withholding electricity at times of peak demand.

How did the price of electricity change when private enterprise models were introduced? Unfortunately, it is not easy to determine the effects of deregulation on price. This is because prices have been influenced by other factors, such as the type of power and the costs of new investments. In the US some states retained a regulated market, while others introduced deregulation laws. Rapid price increases sometimes led to the introduction of rate caps. It is an interesting fact that between 2002 and 2006, prices increased faster in states with market pricing: "While average prices rose 21% in regulated states from 2002 to 2006, they leapt 36% in deregulated states where rate caps expired" (Davidson 2007).

Canadian electricity rates—both residential and from large power consumers—are generally lower than in the US, but not uniformly. One factor that explains the lower prices in Canada is the greater quantity of hydroelectric power generation. The provinces with large volumes of hydroelectricity (Quebec, Manitoba, BC, Newfoundland and Labrador) also tend to have the cities with the cheapest electricity. High rates in Charlottetown are strongly influenced by the price of oil. In some cities the prices of unregulated electricity actually dropped, but the same thing also happened in several regulated ones. One notable fact is that the largest consumers (in other words, big businesses) paid lower rates, although these rates increased faster than the rates given to residential customers.

Choynowski (2004) summarized the pricing record of countries that experienced deregulated electricity markets—in others words, those countries that went from a government-owned model to one based on private enterprise. In terms of supply, the outcome in Australia following the 1990s reform was positive: there were improvements in terms of the supply, the supply's security, and the price of electricity. By contrast, over a similar period, New Zealand experienced stable to slightly rising prices. In Norway prices were generally low in a partially deregulated structure. The UK's reform brought gains in efficiency but with a lack of price competition; this resulted in high prices. Because the US does not have a single national system, there was much variation among states and regions in the extent of deregulation. For example, by 1998 California allowed competition in the supply of energy to utilities, but there were regulations placed on the prices the utilities were allowed to charge consumers. This innovation was soon followed by the power crisis of 2000–1. Power-generating companies such as Enron made enormous profits by manipulating supply to raise prices during times of high demand, but financial crises resulted for retailers who had to sell at controlled prices. Some utilities declared bankruptcy. Widespread blackouts following, forcing California to retreat from this system (Joskow 2001).

Overall, the evidence on pricing shows wide variation among markets. Advocates can argue that instances of failed objectives were the result of market conditions not having been being fully established. Sceptics argue that profit strategies will inevitably lead to market domination and higher prices, which some residents and businesses will be unable to pay (or pay only with considerable difficulty).

Another pro-market argument is that large amounts of private capital are needed to invest in new plants and transmission lines—replacing old equipment and meeting new demands. Some politicians may retreat from supporting expensive investments that would result in higher taxes or energy bills. Conversely, private investors may not want to invest if they see state interference potentially undermining their profit. If this were the case, then encouraging markets would be the best way to ensure the right investment. But the reality is that it seems equally likely that governments will be pressured to pay money to support these projects and still have no real control. Interruptions to supply are at the very least irritating, and often potentially serious. More importantly, as the consumption of electricity goes up, the capacity to produce enough of it may be stretched. This will be the case particularly if aging generation equipment requires replacement. Deregulation is said by supporters to create a system that responds to demand, and with more reliable technological development. In an ideal world, this makes sense.

Environmental Issues

Depending on the kind of fuel and the technology, the generation of electricity may have a range of impacts on the environment. The industry as a whole contributes significantly to air pollution in Canada (Cohen 2006). The worst offenders are coal-fired stations without modern scrubbers. Natural gas plants generate less carbon dioxide per unit than do oil plants. Hydroelectricity, on which much of Canada's electricity depends, is clean as a production process, but the construction of large dams comes at the cost of massive environmental destruction. We saw earlier that nuclear power is also environmentally controversial.

The environmental impact of energy generation is a factor in the operation of the market. Clearly, if a market strategy led automatically to the replacement of old polluting equipment with new, more environmentally friendly technology, it would be good for the environment. However, this is not happening.

Customers facing rates that vary throughout the day are likely to adjust their behaviour in order to take advantage of off-peak rates (Dewees 2005). This should reduce the need for extra equipment to meet peak-time demand. Some types of businesses may find changing usage behaviour easier than most residential customers, whose major usage is fixed by daylight hours or by the necessary timing of household tasks (such as meal preparation). However, washers and driers can be operated at off-peak times. Generally, conservation in heating and air-conditioning—along with other strategies such as improved house insulation—should limit total energy demands, and limit costs to consumers. This should be the same regardless of a public or a market system.

Is the market necessary to achieve these positive changes? Probably not. Some environmental groups, however, have pushed for deregulation and restructuring because they found too little support from governments and regulatory agencies. They advocated instead a market environmentalism approach: here, consumers bring about change by purchasing only those products deemed environmentally (and sometimes socially) acceptable, according to certain criteria. This approach was supported by businesses that normally favour self-regulation (Bailey et al. 2003; Taylor 2005).

Cohen (2006) demonstrates that deregulation has resulted in more coal use. Coal, as we have seen, is more readily available and cheaper than cleaner fuels. Alberta—which has abundant supplies of coal—still produces roughly 50 percent of its electricity from coal (Alberta 2009a). In the US, coal-fired plants that escaped effective regulation under the Bush administration may experience a tighter, pro-environment framework after a review announced in 2009 (Doggett 2009).

Emissions trading may have a modest positive impact. It has the effect of requiring high emitters to purchase discharge rights from those who are lower-level polluters. This increases costs for high polluters and provides incentive to improve their standards. However, if the economics of the situation dictate business as usual, this strategy does not reduce the total pollution. In other words, if the cost of purchasing credits is less than the costs of reducing emissions, there is no incentive to change.

Canadian Provincial Examples

Canada's nuclear energy industry, unlike other sources of power generation, is regulated by the federal government through the Canadian Nuclear Energy Commission, which sets standards for safety, conducts environmental impact assessments, and issues production licences. In other respects, the production and distribution of electricity is a provincial responsibility except that the federal government controls international trade.

All of us have an interest in clean air and reducing the negative impacts that come with rapid climate change. However, this does not mean that the choice of energy can be based on these criteria alone. Nuclear energy, for example, has a low environmental impact in the short term, provided that there are no serious accidents like Chernobyl (see p. 95). The long-term problem of disposing of radioactive waste remains unsolved. The production costs of nuclear energy are also high. Many parts of Canada rely on hydroelectric power, which can be produced at a low average cost and is a clean energy source. But hydroelectric power is based on dams that are destructive to the land around them and reduce the flow of rivers lower down their courses. The interests of those affected by dam sites have often been sacrificed in the interests of those who live far away and consume the energy. Coal is still abundant and cheap, but generates serious pollution, even with the latest technology. As oil and natural gas become scarcer, their costs will rise and less damaging renewable resources will go down in relative terms. For now, however, they are more expensive and unlikely to be widely adopted without state intervention in the market and higher costs for consumers. Clearly there are many difficult decisions around electricity. Given Canada's regional variation with respect to energy resources, it is no surprise that the provinces take different approaches. I discuss these differences in Canada's most populated provinces below.

It should be noted that there is relatively little transfer of electricity among the provinces (with some exceptions, such as the arrangements between Alberta and BC). The provinces focus on their own needs and are more likely to sell any surplus to the US than to neighbouring provinces (Cohen 2002). Under pressure both to ensure adequate investments for the future, and to sell electricity on the open market, we noted above that in the 1990s several Canadian provinces moved toward privatization.

Ontario

Ontario depended almost entirely on hydroelectric sources of electricity to power its urban industrial growth until that source became exhausted in the 1950s. Coal-fired plants were introduced, and these were soon followed by nuclear generators. By the early 1990s, 20 reactors operated at three sites—Pickering, Bruce, and Darlington.

By the time it came online, the cost of the Darlington plant had created a large debt load for Ontario Hydro. This debt load coincided with lower demand for electricity in 1992–93. Cutbacks were made. Ontario Hydro itself recommended a market approach to electricity. In 1995 the Conservative government set in motion a review process that led to the market experiment of 2002. The electricity market opened in May; by July, average prices had approximately doubled. There was a great deal of public protest, which led to a quick retreat later in the year. The government capped transmission and distribution charges, and introduced a rebate to consumers for the amount that the price exceeded 4.3 cents per kilowatt hour (Ontario 2002; Swift and Stewart 2004). This intervention by government created a good deal of uncertainty, scaring away potential investors who feared loss of profit.

By the mid-1990s, the oldest nuclear reactors were aging, inadequate, and in need of replacement. In 1997 the Ontario government supported an Ontario Hydro plan to close the seven oldest reactors and replace lost capacity with coal-fired plants— in spite of the greenhouse gas emissions produced by coal. As awareness of the environmental and health impacts of coal grew, a commitment was made in 2003 to phase out coal-based electricity. At the same time, however, the blackout in the summer of 2003[1] instigated action to address the problem of inadequate infrastructure. The following year, the Electricity Restructuring Act created the Ontario Power Authority (OPA), a body designed to plan for long-term investment and stability.

The OPA proposed a new plan in 2005. In this plan, a revitalized nuclear energy sector would remain the most important single source of electricity. Coal-fired production would be eliminated. Natural gas and renewable sources would make up the difference (OPA 2005). The plan foresaw a radical expansion of power from renewable resources, but key weaknesses were continued reliance on nuclear energy and insufficient attention to reduced consumption. The Ontario government advised the OPA to formulate a new plan that would address these concerns. (In terms of nuclear-based generation, however, the government required only that it be restrained modestly to no more than the status quo—roughly 50 percent of electricity from nuclear plants.)

The outcome remained unclear for several years. In 2007 the OPA submitted its Integrated Power System Plan for the Ontario Energy Board's approval. But hearings for the Plan were adjourned indefinitely on 2 October 2008, pending the OPA's response to a supplementary directive from the Ministry of Energy in September 2008. This new directive required the OPA to revise various targets and to consult more extensively with Aboriginal peoples (OEB 2009). To an extent, this review process was superseded by a parallel focus on green energy.

Notes on Electricity in Other Provinces

In 2001 Alberta was the first Canadian province to introduce an electricity market for consumers. Like Ontario, this led to large price rises followed by a rebate that probably inhibited conservation measures on the part of consumers. Recent statistics show that more than three-quarters of Albertans chose a regulated rate from a local utility over entering into separate term contracts with private utilities (Alberta 2009a). A key problem for Alberta is the expansion of its infrastructure to keep up with growing demand.

Cohen (2002) has argued that provinces exporting electricity to the US have effectively been pressured to introduce market competitiveness, at least in terms of the generation of electricity. This process has seen minimal impacts in provinces like Manitoba, however, which can provide electricity at low cost from its state-owned hydroelectricity company. Newfoundland and Labrador (except indirectly via Quebec), Nova Scotia, and Prince Edward Island export no electricity and have no plans to deregulate.

The BC Power Commission was formed in 1945 to extend electricity throughout the province. Started in 1898 as a private company in Victoria, BC Electric was bought by Power Corporation in 1928. In 1961 the provincial government purchased this company and in 1962 the two amalgamated to form BC Hydro. BC Hydro was restructured 40 years later into three business entities concerned with production, transmission, and distribution (see the BC Hydro history website for full details). BC Hydro still functions as a state agency that reports to the government and is regulated by the British Columbia Utilities Commission. The focus of BC policy is on hydro and renewable energies with continued public ownership and the rejection of nuclear energy (British Columbia 2007).

The following year Ontario passed its Green Energy and Green Economy Act, which pushed the province to the forefront in North America with regard to electricity production and conservation. The core objectives include measures to increase the production of power from renewable sources by 25,000 megawatts in 2025 compared with the 2003 level (Ontario 2009). This represents about two-thirds of total requirements for that year as estimated in the OPA (2005) report. The target also assumes a successful conservation program to complement the changes in production.

Based on the model of Germany's feed-in tariffs, Ontario will periodically establish the prices that various "green" producers will receive and guarantee that these producers will have access to the grid. Prior to this Act, the lack of such support had seriously impeded the growth of large-scale renewable energy projects. Project developers should experience a smoother application and approval process, while consumers will receive incentives to conserve energy. It is expected that 50,000 jobs will result from expanded green energy in the first three years alone. Overall, this Act appears to serve all interests well and has received widespread initial support. The Ontario public, in fact, was 87 percent behind the Act just before it was passed (Green Energy Act Alliance 2009).

Quebec

Hydro-Québec was formed as a state company in 1944 as a result of taking over the Montreal Light, Heat and Power Company. The new company was solidified in the full nationalization of 11 remaining producers in 1963, and now Hydro-Québec is the country's largest producer of electricity. The company manages 58 hydroelectric plants, one nuclear plant, four thermal generating stations, and one wind farm (Hydro-Québec 2009). It remains a regulated state service in which over 95 percent of its electricity comes from hydro sources. Hydro-Québec is obliged to sell up to 165 terawatt-hours of energy at a fixed rate of 2.79 cents per kilowatt-hour to the provincial distribution and retail organizations. This sale takes place in accordance with the "patrimonial" or heritage commitment to provide the population with a stable supply of inexpensive electricity. This amounted to about 96 percent of the electricity that was required in 2008. Part of Hydro-Québec's production is sold to grids in the northeast US and part is supplied from the Upper Churchill in Labrador on a long-term contract. Because the regulated heritage price is so low, the small amount exported (8 percent) at market prices accounted for a whopping 46 percent of net income in 2008 (Hydro-Québec 2009).

As an exporter of cheaply produced electricity, security of supply is not an issue in Quebec (except that the transmission lines have been susceptible to collapse under severe weather conditions). Nor is privatization on the cards, despite some claims that Hydro-Québec is less efficient than a private operator would be. Since 2000, Hydro-Québec has functioned as three separate companies in terms of production, transmission, and distribution. All three, however, are public enterprises subject to the overall regulation of their activities (Cohen 2002). No Quebec government has indicated any plan to introduce a competitive market for electricity at the retail level and none is in the works (Lessard 2009).

Quebec's reliance on hydropower ensures the clean processing of electricity, but this takes place at the cost of habitat destruction when large dams are created. The damming of northern rivers has proven controversial, especially in the James Bay area where the Cree First Nation fought a protracted battle to protect their traditional hunting grounds. Following a legal injunction against further development in 1973, the Quebec government was forced to reach a compensation agreement with the Cree in 1975. Later developments also proved problematic. Another large-scale development, the Great Whale project, was finally shelved in 1994. This followed years of struggle by Cree and environmental groups in Canada and the US, to which much of the new power was to be sold (Hornig 1999).

As with most other provinces, Quebec encourages consumers to conserve and promotes renewable energy. However, continued low prices from the dominant hydro source create little financial incentive to change habits. For example, a plan to introduce "smart" metres that would have allowed differential charges based on time of consumption were abandoned in light of the small changes that might be expected; this contrasts with a successful introduction in Ottawa (CBC News 2007c). On the other hand, some initiatives have worked. For example, in 2008–9, an offer to buy old refrigerators for $60 removed 83,960 from use (Lalond 2009). Of course, given their average age of 25 years, most would probably have been scrapped soon in any case.

Ontario and Quebec, like other provinces, continue to search for the most efficient and environmentally friendly ways to manage electrical distribution. Globally and within Canada there has been wide variation in the extent to which the electricity industry has been deregulated. Deregulation does not ensure lower prices or guarantee the investments needed for a secure supply. Much of Canada has the advantage of cheap hydroelectricity, but other renewable energy sources are needed to replace older coal-based plants and to support future increases in consumption. Chapter 7 considers renewable energy in more detail.

Chapter 7

Alternative Energy

We have seen that alternative and renewable energy currently form a small part of our energy mix. However, in the face of climate change and a looming energy crisis, these other sources have potential applications.

Nuclear Energy

Nuclear energy is contentious. The nuclear alternative for generating electricity is relatively clean burning, but controversial because of disposal problems with by-products and the potentially drastic impact from a major accident (however unlikely that may be). The US is the world's largest producer, and France depends most on nuclear energy for its supply of electricity (78 percent). At present, Ontario is the centre of the industry in Canada with 20 reactors, but there are also reactors in Quebec (one) and New Brunswick (one) (Canada 2009d).

Uranium is the mineral that nuclear power plants require at present, although in future we may find the more abundant thorium introduced in new facilities or to CANDU reactors produced in Canada. The known reserves of uranium will last about 80 years at current rates of use (World Nuclear Association 2009). This suggests that rapid expansion of uranium-fed reactors can be undertaken only if large new deposits are discovered.

All reactors to date are based on the process of nuclear fission. In a specially designed reactor, the nucleus of a uranium atom is struck by a smaller particle. The uranium nucleus splits into two parts, releasing energy. Also released in this collision are neutrons, which go on to collide with other uranium atoms, and a self-sustaining chain reaction takes place. The heat generated from these reactions is harnessed to convert water to steam, which drives turbines to generate electricity. (Under some very specific conditions, uncontrolled chain reactions have been used to produce nuclear explosions.)

Although building reactors is expensive, nuclear energy is cheap to produce. Moreover, even allowing for its various inputs, it contributes very little to greenhouse gas accumulations. However, refurbishing and new construction is so expensive that nuclear energy is rapidly losing any cost advantage over renewable sources. Despite its positive features, the problems are serious. While the probability of major calamities is low, the potential harm is great.

Nuclear fission is capable of generating large amounts of energy compared with conventional fuel cells, but its products are radioactive. Although uranium provides clean electricity, the process of converting the uranium to electricity through nuclear reactors generates radioactive waste that is dangerous in the long term and requires careful disposal. High-level radioactive waste takes about ten times its half-life (the point where it has lost half its initial level of radioactivity) to reach a safe level for humans. For plutonium-239, which is the most dangerous element, this amounts to about 240,000 years (IEER 2005; US NRC 2003).

Accidents are rare, but potentially disastrous, as at Chernobyl (Ukraine) in 1986. Although there were 28 deaths from radiation sickness in 1986, thousands received doses that were likely to produce fatal cancers. Scientists cannot be precise about the impact of low doses of radiation, and there is disagreement about how many deaths should be attributed to Chernobyl. A World Health Organization (WHO) expert investigation estimated that the radiation from Chernobyl would cause about 4,000 deaths in the area most at risk and roughly 5,000 additional deaths among those further from the source in Ukraine, Russia, and Belarus (Bielo 2009; WHO 2006). Other reactors have experienced various problems that have been contained before causing mass damage, as in the widely reported partial meltdown at Three Mile Island, Pennsylvania. More recently, a close call occurred in 2002 at the Davis-Besse plant on the shore of Lake Erie. When the plant was shut down inspectors found that boric acid had eaten through a lid that retained radioactive water inside the reactor. Nuclear steam was held back only by a thin steel liner that was bulging under stress when discovered (Bielo 2009). This problematic situation had escaped the attention of at least two inspections. This reactor was closed for two years.

Finally, the possibility of terrorist action against nuclear facilities is frightening to some people. It seems impossible to guard against all the ways that a terrorist group might seriously damage a nuclear plant or obtain materials from the plant that could be used in the manufacture of nuclear weapons. Moreover, there are numerous cases in which potentially dangerous materials are stored under security protection that is modest at best (for example, in South Africa where a nuclear facility was easily breached in 2007). In some cases, like Pakistan, even if security is strong there remains a serious threat (Bunn 2008). Electricity by other means is preferable to many.

Is Nuclear Energy Safe? The Case of EDF

The Three Mile Island nuclear power plant in Pennsylvania is well known as the site of a serious incident in 1979 that led to a large-scale evacuation. In 2009 another leak exposed 20 workers to radiation. Reporting on this event, physician Gwen DuBois (2009) expressed particular concern about plans to build a new plant close to Baltimore in which Electricité de France (EDF), the giant French utility that is 85 percent state-owned, would be a 50 percent shareholder.

This company's safety record in Europe is suspect. For example, in October 2009, more than 1,500 tons of "really dirty" spent fuel were found in cans in Seversk, Siberia. This waste lies in the open air and is visible from satellites. EDF claimed that it only shipped uranium that can be recycled and that what happened to it was essentially the responsibility of the Russians (Samuel 2009). The point is that it is difficult to deal safely with by-products of nuclear energy, and large-scale producers like France appear to be exporting their problems.

A further issue is that some nuclear power companies are overextended financially, which raises concern about their ability to ensure that their facilities are safe and reliable. Again, EDF is a prime example. Its French plants are in need of major overhauls to improve reliability and its expensive international expansion has loaded the company with debt. An official complained that, "Every morning we get a new safety request" (*The Economist* 2009). It must be tempting in a difficult financial environment to minimize safety.

Renewable Energy

Transportation

Alternative transportation energy has been developing quickly in several areas. Emission controls of conventionally fuelled motors have become much more effective, including "clean" diesel or diesel-hybrid engines from Volkswagen and Peugeot that give superior distance per litre. Many automobile companies (Toyota, Honda, and Ford are leaders) have already introduced hybrid vehicles that function on a combination of conventional gasoline motors and electrical batteries. This system reduces fuel consumption and carbon emissions by about 50 percent (Maslin 2009). Vehicles fully powered only by electrical motors may become more common (the Chevrolet Volt is scheduled for 2010), but they will surely increase pressure on electricity supplies. The final outcome will only be positive if the electricity source is renewable and carbon-free.

Fuel from biological sources, such as ethanol, can be mixed with gasoline to power vehicles. As we have seen in recent years, however, the rapid expansion of this option raises issues of land use and its impact on food supply. It seems likely that it will adversely impact the living costs of poorer people. Biofuels are nevertheless part of the solution to moving away from fossil fuels. Indeed, Mathews (2007) proposes that non-OECD countries could follow Brazil's example and generate enough ethanol to cover 20 percent of the requirements of industrial countries by 2020. He argues that low-grade, underutilized lands in the South that could be converted to crops suitable for ethanol are more than enough to meet this need. In 2008 the US and Brazil were the world leaders in production (Table 7-1), which has been rising rapidly.

Hydrogen-powered vehicles are another possibility. Several manufacturers have produced examples ranging from a small Honda sedan to a luxury BMW prototype and a Mercedes bus. For various reasons, these alternatives have not been widely adopted. Hydrogen is much more volatile than gasoline—in other words, it is highly flammable—so users must be confident in the safety features of hydrogen fuel cell vehicles. It is also difficult to find refuelling stations for these vehicles at present, a fact which inhibits their usage. Probably the biggest problem is that the separation of hydrogen from oxygen requires electrical energy; any substitution of clean burning hydrogen for oil will put pressure on electricity

Energy Companies and Renewable Energy

What is the position of global energy companies on renewable energy? With the exception of ExxonMobil, all major oil and gas companies have directed substantial research and even investment funds to renewable energies, although the amounts are small compared with what they spend on oil and gas.

For some years (2000–7), BP sought to rebrand itself as Beyond Petroleum and became the world's largest producer of solar power equipment. In the global recession of 2008–9, however, the majors have retrenched, demonstrating that they are first and foremost oil and gas companies. In effect, they have rejected the Obama administration's desire for $150 billion to be invested in renewable energy over ten years (Mouawad 2009b). BP and Shell withdrew from major wind power developments in the North Sea. BP planned to sell off its alternative power division, valued between $5–7 billion (Hotten 2008), but eventually decided against this. By 2009 Shell was restricting its attention to biofuels. Total SA had turned to tidal and nuclear power. ExxonMobil continued along its path of minimal focus on alternatives, claiming that conventional oil and gas remained even more necessary (and more profitable) than alternatives. Yet, by 2009, this company announced investments in an electric car, unconventional shale gas extraction, and algae biofuel (Howell 2009).

Some major energy companies in Canada (Enbridge, Suncor, Nexen, and TransAlta) contribute to the renewable energy sector. In 2000, Enbridge—a company primarily known for its oil and gas pipelines—entered into a joint project with Renewable Energy Services Canada to build a new wind farm in Southwest Ontario. This farm would be capable of providing energy to 33,000 homes (Newnet 2009). With involvement in five wind projects and one solar project (with a total capacity of 380 megawatts), as well as a hybrid fuel-cell power plant, this company is looking to the future. (More information is available on the company's website.) Suncor, the oil sands plant operator, also invests in renewable energy. It cut back on renewable in 2009, cancelling expansion of its ethanol production. In all cases, these developments so far are modest adjuncts to their main businesses of producing and transporting conventional energy.

supplies from alternative sources. Moreover, compressing hydrogen for storage also requires large amounts of energy.

In principle, hydrogen-powered vehicles based on green sources of electricity could become widely adopted by consumers, but the transportation industry has been slow to move in this direction (Black 2006). Honda was the first to make its hydrogen vehicle—

Table 7-1 World Ethanol Production, Millions of Gallons

Country	2007	2008
USA	6,499	9,000
Brazil	5,019	6,472
Europe	570	734
China	486	502
Canada	211	238
Thailand	79	90
Colombia	75	79
India	53	66
Australia	26	26
Other	82	128
World	13,101	17,335

Source: United States 2009

the FCX Clarity—available to the California public in 2008 (Lampton 2009). Rather than use green electricity to generate the hydrogen and refit countless fuel stations, most makers see hybrids and electric cars as the path of the future.

An interesting innovation in BC is the establishment of the "hydrogen highway." This concept involves hydrogen fuel stations in Victoria, Vancouver, and north to Whistler. This demonstration project will operate a fleet of hydrogen-powered buses.

The Hydrogen Highway is targeted for full implementation by 2010. Canadian hydrogen and fuel cell companies have invested over $1 billion over the last five years, most of that in BC. A federal-provincial partnership will be investing $89 million for fuelling stations and the world's first fleet of 20 fuel cell buses (British Columbia 2007).

Electricity

Apart from nuclear energy, the alternatives to fossil fuels for electricity generation consist of different types of renewable energy. I will summarize these here briefly before considering the situation in Canada.

Falling or flowing water is the basis of hydroelectricity, which is by far the world's largest source of renewable electricity at present. It is particularly important in Canada. Once constructed—usually a long and costly process—hydroelectricity is cheap to produce and generates little in the way of greenhouse gases. We have noted above, however, that large dams create serious habitat destruction and may also displace human populations. Downriver flows are affected in the long term and this may cause political disputes as well as environmental problems. The occasional dam failure has produced massive damage and loss of life. Dams are also potentially good terrorist targets.

In addition to dams that are currently under construction, it is unlikely that much additional large-scale hydro development will take place. Some countries, however, still contain untapped resources; this is the case with Canada. In the future, we may see primarily smaller-scale hydroelectric plants. Some of these projects do not depend on dams, but rely instead on river flow to turn turbines (this electricity is often called run-of-the-river hydro). It is particularly suitable for more remote areas.

Solar energy arrives on earth as high intensity radiation, which changes into heat whenever it strikes a surface. Solar heat collectors trap solar energy as heat. From these collectors, the heat can be harnessed to different applications. Solar photovoltaic collectors transform sunlight into D/C current; the amount of power that becomes available depends on the number of cells or modules that are linked together. Germany is the leading country for solar energy—this is notable in view of the popular belief that only the sunniest locations are suitable. (It is true that solar power is more readily available and better-suited to sunny locations.) Initial costs are high, but newer technologies may reduce these costs significantly. Once installed, solar systems only have routine maintenance costs, given that sunshine is free. Like wind power, the intermittent nature of solar energy means that it must be combined with other sources (Pembina Institute 2007).

Geothermal energy comes from hot, molten rock close to the surface of the earth, as in Iceland. Pumping water into this rock heats the water and generates steam. Geothermal energy is limited by location. Even so, however, all buildings can be partially heated by water that is pumped into underground boreholes (Maslin 2009). Overall, geothermal processes are expanding rapidly and making a real contribution to renewable energy production. Five percent of California's total electricity is produced this way. In the first half of 2008, "total world installed geothermal power capacity passed 10,000 megawatts and now produces enough electricity to meet the needs of 60 million people" (Dorn 2008).

Wind has enormous potential, but again it is restricted by geography. In terms of reliability, the fact that it is an intermittent source means that it can primarily supplement rather than replace other forms. However, existing turbines are active about 80 percent of the time and, if located in different regions that are connected by a grid, turbines can provide a more stable supply of electricity. In Denmark, government support has allowed wind turbines to become much larger and more efficient generators since the 1990s. World capacity has grown rapidly with the US and Germany at the top of the list by 2008 (Figures 7-1 and 7-2).

Wind farms are considered unattractive by some, and if not well sited may harm bird life. Clearly these are less serious drawbacks than most other forms of energy. Indeed, properly sited wind turbines have little impact on wildlife and are supported by the Audubon Society (Pembina Institute 2007). Local residents may object to wind farms (even though they create employment). One study showed, for example, that local residents expected serious negative consequences from a development off Cape Cod, Massachusetts; but these expectations were completely inconsistent with the impact assessment study. A different study found that younger, better educated people are more likely to be supportive (Firestone and Kempton 2007). Researchers in Europe also found that local resistance could be addressed by encouraging ownership and creating institutions to promote local participation rather than imposing the development from above (Breukers and Wolsink 2007; Joberta et al. 2007).

Electricity and heat from burning biowaste of various types is becoming important in many countries, harnessing the by-products of wood processing and other waste. Sweden is the leading country in Europe to make use of biowaste (Lundeberg 2009), with an ambitious program for the future. In 2008 biomass accounted for 5.2 percent of Sweden's electricity supply (European Commission 2008).

Tidal and wave power uses the natural movements of the ocean to generate electricity. This approach is still in its infancy and may generate serious habitat damage in coastal areas (Maslin 2009). However, if developed with an attention to the local environment, there is abundant energy in the ocean.

Overall, renewable energy in the form of electrical power generated from geothermal, solar, wind, wood, and waste sources increased by about 250 percent from 1989 to 2006 (Figure 7-3).

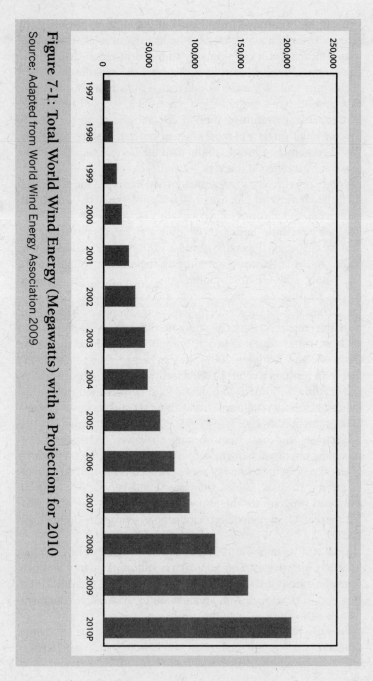

Figure 7-1: Total World Wind Energy (Megawatts) with a Projection for 2010

Source: Adapted from World Wind Energy Association 2009

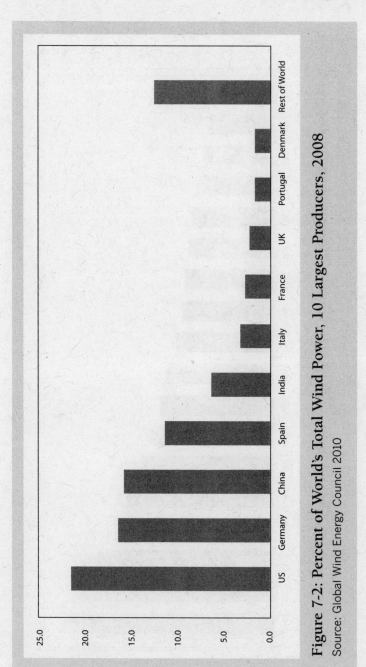

Figure 7-2: Percent of World's Total Wind Power, 10 Largest Producers, 2008

Source: Global Wind Energy Council 2010

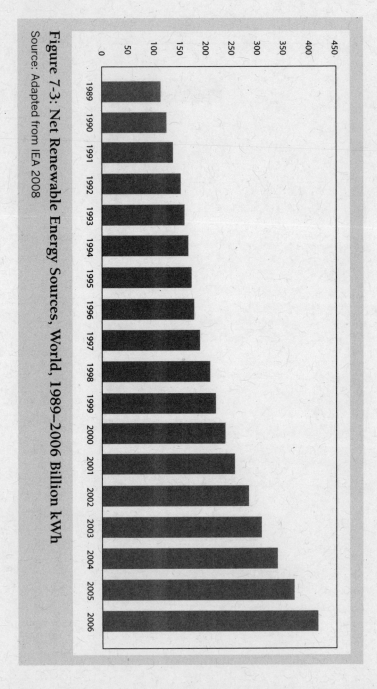

Figure 7-3: Net Renewable Energy Sources, World, 1989–2006 Billion kWh

Source: Adapted from IEA 2008

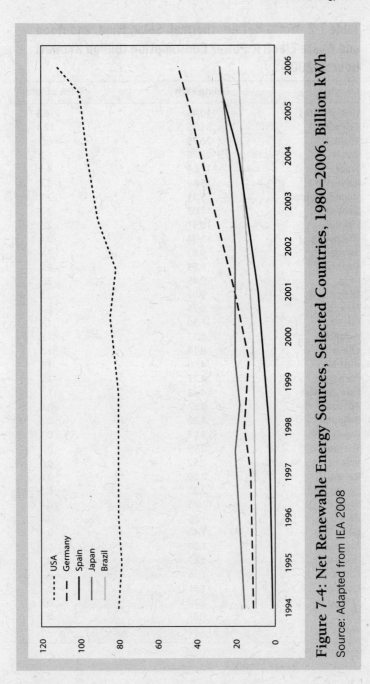

Figure 7-4: Net Renewable Energy Sources, Selected Countries, 1980–2006, Billion kWh

Source: Adapted from IEA 2008

Table 7-2 World Net Geothermal, Solar, Wind, and Wood and Waste Electric Power Consumption (Billion Kilowatt Hours), 2006

Country	Billion kWh	% of Total
United States	110.40	26.6
Germany	51.55	12.4
Spain	26.75	6.5
Japan	25.86	6.2
Brazil	17.09	4.1
Italy	15.49	3.7
United Kingdom	15.34	3.7
Canada	11.03	2.7
Finland	10.51	2.5
Philippines	9.94	2.4
Sweden	9.83	2.4
Denmark	9.53	2.3
India	9.45	2.3
Netherlands	9.07	2.2
Mexico	8.75	2.1
France	7.30	1.8
Indonesia	6.33	1.5
China	6.16	1.5
Austria	4.90	1.2
Portugal	4.77	1.2
New Zealand	4.48	1.1
Australia	3.57	0.9
Belgium	3.30	0.8
Russia	3.05	0.7
Thailand	2.99	0.7
Poland	2.52	0.6
Iceland	2.50	0.6
Switzerland	2.25	0.5
Greece	1.75	0.4
Ireland	1.66	0.4
Costa Rica	1.48	0.4
Argentina	1.39	0.3
Hungary	1.37	0.3
Kenya	1.16	0.3
Norway	1.11	0.3
El Salvador	1.10	0.3
Chile	1.07	0.3
World Total	414.31	

Source: EIA 2008

While this looks dramatic, renewable sources accounted for only 2.6 percent of global electricity in 2007 (IEA 2009).

Although the US has always been the largest single producer of electricity from renewable sources, some countries, especially Spain and Germany, exhibit more dramatic increases in recent years (Figure 7-4). By 2006, excluding hydroelectricity, Canada stood in eighth position in the world on other alternative energies (Table 7-2). On a per capita basis, some smaller European countries, notably Denmark, stand out in comparison with others.

Canada and Renewable Energy

The Canadian public is generally supportive of renewable energy. In 2008 a national survey reported that 65 percent were willing to pay more for renewable energy and 88 percent felt its development should be mandatory. Almost as many respondents took the view that it was good use of taxpayers' money (CanWEA 2008). Although Canada is reasonably placed in terms of resources for developing renewable energy, it is failing to take advantage of new opportunities. Investment is flowing into countries like Germany, which have made important commitments in key areas (Castaldo 2008).

Canada is at the forefront in one area of renewable energy, ranking second in the world in hydroelectricity. Canada's hydroelectric power expanded rapidly in the early twentieth century and continues to supply about 60 percent of the country's electricity, although the high capacity sites near large population centres have all been tapped. Hydroelectric energy is developed on a provincial basis. With much inter-governmental conflict, no national grid has been established in Canada, and economic benefits have been lost (Froschauer 1999).

The federal government's rhetoric certainly supports renewable energy, but programs to provide stimulus are modest at best. The main support program (ecoENERGY for Renewable Power) ran out of money in 2009, and there is no sign of it being extended in 2010. This program had provided a subsidy of one cent per kilowatt-hour, which was small but helpful. Meanwhile, the US is outspending Canada at a 14:1 ratio (Pembina Institute 2009).

The provinces also play an important role in energy development and have been supportive of renewable energy in some critical areas. Most important, Ontario's new Green Energy Act is likely to provide a real stimulus to the industry (see Chapter 6). Even Alberta's recent energy strategy outlines plans to take part in the long-term substitution of renewable sources for oil, gas, and coal. As oil becomes more expensive to produce, Alberta recognises that alternatives are desirable (Alberta 2008). However, it is worth noting that given the pace and promotion of oil sands development, these words ring hollow to many ears.

British Columbia has moved forward with an energy plan that includes an Innovative Clean Energy Fund of $25 million per annum to support new clean technologies and alternative fuels. As part of its plan to become energy self-sufficient by 2016, BC expects to expand its biofuels sector, including the use of timber infested by pine beetles. The province only purchases hybrid vehicles. British Columbia is also increasing the mandatory level of ethanol and introducing a 5 percent renewable sources standard for gasoline and diesel fuel (British Columbia 2007). The hydrogen highway, mentioned above, is also part of this plan.

With regard to the new renewable energies, Canada has performed unevenly and below its potential. This is partly explained by the relatively low cost of electricity and the lack of concern about energy security in recent years. Solar energy investments remain low in Canada despite suitable conditions in much of the country. A forward-looking business, ARISE Technologies, is an interesting case in point. Located in Waterloo, Ontario, ARISE manufactures its solar cells—capable of generating 35 megawatts of power annually—in Germany, where financial incentives and a supportive network are much more attractive than in Canada. And this is only one of many cases (Castaldo 2008). Geothermal energy has also been slow to emerge in Canada, with too little support from the state to assist with the long time frame required for development.

Ethanol and other sources of energy from crops or biowaste have been more successfully developed in Canada. British Columbia is the leading centre with some 50 percent of Canada's capacity. Much of this capacity takes place in co-generation, in which the forest products industry generates heat and electricity from its own waste. In 2009, BC reported 800 megawatts of electrical capacity from biomass sources—enough to heat 860,000 households (British Columbia 2009). "The British Columbia wood pellet industry also enjoys a one-sixth share of the growing European Union market

Who is Pushing for Change?

For about 30 years many scientists, environmentalists, and other less directly involved citizens have been pressing states, international organizations, and companies to respond positively to the need to improve our energy usage (e.g., Brown 2009; EREC and Greenpeace 2008; IPCC 2007). As these groups urge research into and development of renewable sources of energy, they add to economic pressure on producers (such as corporations operating in Canada) to reduce emissions associated with extraction, and on consumers to reduce consumption. The environmental movement is also a major factor pushing states towards support for alternative, preferably renewable energies, and encouraging companies to develop these energy sources.

for bioenergy feedstock" (British Columbia 2007). Existing and new generation biofuels are under development with a significant contribution possible in the struggle to control harmful emissions (see the summary in Canada 2008b). The federal government requires that commercial fuels include at least 5 percent renewable sources in 2010 and is investing $1.5 billion between 2007 and 2016 to promote this change.

Wind power is the most developed and cost-efficient of Canada's new renewable energies. Across the country, in different locations, it is now possible to see giant turbines whirring in the wind. Their contribution to electricity is growing rapidly as indicated by the jump from 137 megawatts of installed capacity in 2000 to 3,249 megawatts in 2009 (CanWEA 2009). Nevertheless, wind supplies only about one percent of Canada's electricity. Canada does not make the top ten in this category (Figure 7-2) and the rapid expansion is only on par with the global trend (WWEA 2010). The national wind energy association hopes that winds will power 20 percent of the country's electricity by 2025. The association expects billions of private investment in manufacturing and services, but stimulus from governments would still be needed to meet this target.

Overall, Canada's provinces are pushing forward with various plans for adding new renewable sources of electricity, while maintaining hydro and cutting back on fossil fuels. Some progress is also evident in transportation fuel. It is difficult to predict the amount of change we will see in another 20 years, but public support is generally strong for renewable energy and many are willing to pay more to protect the environment. Rising oil prices are likely to speed up the transition, but greater financial incentives from the federal government would help ensure that Canada keeps pace with other industrial societies and resists becoming overdependent on fossil fuels.

Chapter 8

Conclusion

Automobility is a concept introduce by John Urry (2008) to describe a core aspect of the contemporary world. The term conveys an assemblage of vehicles, drivers, roads, service stations, petroleum, control regulations, driving aids, certification processes, and other related technologies that form a complex self-expanding system embodying a particular culture or way of life. However, this system will not function without the necessary physical resources and petroleum supplies—all of which are becoming increasingly scarce. The theory of complex systems suggests that pressures for change will reach a tipping point after which a newly structured assemblage may emerge.

Urry asks whether a post-car system of transportation is possible. He sees a number of small changes that might generate such an outcome if they took place in the correct order. We have seen that new fuels—biofuels and hydrogen—are being developed. In the future, lighter materials and more recycling are strategies designed to reduce energy consumption without sacrificing economic strength. Vehicles may be increasingly controlled by software rather than hardware. Car-sharing and other forms of access—rather than just private ownership—are becoming more important. Overall transportation governance may lead to close integration of different technologies. Urry expects the coming decades to produce a new structure for transportation such that the current "steel-and-petroleum" model of automobility will die:

> It will develop in some place and suddenly it will be the fashion. It will probably emerge in a small society or city-state with very dense

informational traffic that can convert into a post-car configuration (Iceland perhaps as an interesting prototype which has already announced itself as the first Hydrogen Society). And it will emerge out of crisis when the high carbon economy/society has to be massively restructured (Urry 2008).

Because the networks that link us to each other and to the environment have the characteristics of complex systems, neither Urry nor I can accurately predict the precise form of the future. Nor could we predict when the critical changes will happen. In this book, I have set out the environmental and social pressures that have become important issues for Canadians. These issues are pushing the society towards major changes. The preceding chapters describe how is this happening, by which means, and (where possible) in whose interests. Energy does matter. We need to understand the choices that lie ahead of us.

Until now, energy security only became an issue at times of externally imposed stress, as in 1974–75. By the 1980s, Canada accepted its role as a committed supplier to the US, regardless of our own domestic needs. After the end of the National Energy Program there appeared to be little government concern about energy security. However, the commitment of energy to the US and a probable slowdown in oil sands extraction (a result of costs and environmental concerns) are likely to make energy security more critical in the coming years.

The concept of peak oil implies that petroleum fuel, the basis of industrialization, cannot sustain industrial growth and lifestyles indefinitely. The problems connected to global warming—with fossil fuel combustion the greatest contributor—are equally compelling. Expanding our search for ever more oil and gas into the fragile environments surrounding the oil sands and offshore rigs also draws issues of social equity to the forefront. Energy cannot, and should not, be pursued blindly. The interconnected needs for secure supplies of affordable energy that do minimal harm to the environment have brought provinces and the federal government, however reluctantly, to regulate markets and investments in a range of different ways.

Canada, along with the broader world of mature capitalism, is undergoing a period of extreme uncertainty and restructuring. This is a result of the twin problems of economic crisis and the environmental impacts of industrial growth. The limits to resource extraction and the scale of environmental degradation generated by human activity threaten the way of life built around dominant global capitalism (Foster 2002; Wallerstein 2004).

Alternative energy is now seriously considered in planning and policy for Canada as well as other societies. Environmentalists promote research into and development of renewable alternatives based on hydrogen fuel cells, wind, solar, biofuels, geothermal, and tidal power. This adds to the economic pressure on producers (such as corporations operating in Canada) to reduce emissions associated with extraction and on consumers to reduce consumption.

Energy is a core part of the way Canadians live and it is in the process of changing. It will not be easy. It may involve a reconsideration of what counts as quality of life. We are already seeing challenges to long-held assumptions about personal transportation and energy consumption. In the future there may be limits to personal movement and how we use space. Even so, we do not need to despair: urban spaces can be reconstructed in ways that encourage walking and cycling, reducing the time spent travelling to and from employment, and radically reducing unhealthy emissions.

There are many ideas and initiatives out there. The processes of transition may be stressful and may create new inequalities that need attention, but change is clearly necessary. To John Urry, this might be possible if the global crisis becomes so severe that a "disaster capitalism" emerges to profit from a new post-car society. Whatever the process, we must look beyond our present problems toward a society that functions successfully within the natural world.

Glossary

Alternative energy
Sources of energy other than fossil fuels.

Annex B countries
Annex B of the Kyoto Protocol includes all developed countries. It also includes countries with economies in transition (such as Poland and Ukraine) that have specific targets for the reduction of emissions of greenhouse gases under the terms of the Protocol.

Annex II countries
Countries included in Annex II to the United Nations Framework Convention on Climate Change, consisting of all the developed countries in the Organization of Economic Co-operation and Development. They are expected to provide financial assistance to developing countries to help these countries meet their reporting obligations as well as to develop appropriate technologies.

Biofuel
Fuel from recently dead biological material, most commonly plants.

Cap-and-trade system
The cap refers to the maximum amount of carbon dioxide that an organization is permitted to emit into the atmosphere in a given time period. An organization that emits less than its allowance may trade or sell the difference to an organization that will emit more than its allowance. Initial permits are set by a central authority and may be reduced over time.

Carbon sequestration
Capture and long-term storage of carbon underground or under the ocean in order to reduce emissions into the atmosphere.

Carbon tax
A tax on the use of fossil fuels, based on their carbon content, that is designed to reduce their usage and consequently the emission of carbon dioxide.

Convention on Climate Change
The United Nations treaty organization formed in 1992 to develop a plan to reduce greenhouse gas emissions and their impact on climate.

Crude oil (or petroleum)
A liquid mixture of hydrocarbons found in underground reservoirs.

Fossil fuel
Fuel (oil, coal, and natural gas) with high carbon and hydrogen content that has been produced by the decomposition of organisms over millions of years.

Giant oil field
Any field with a reservoir expected to produce at least 500 million barrels of oil or natural gas equivalent. The largest fields, containing at least five billion barrels, are called super-giants.

Global Climate Coalition
A once-influential group that claimed human activity was not the main source of global warming, and campaigned against the Kyoto Protocol.

Greenhouse gas
Gases which have been linked to global warming as a result of their increased concentration in the atmosphere where they trap some of the heat (in the form of long wave radiation) that is reflected from the earth and send it back to the lower atmosphere. The main greenhouse gases by volume are water vapour, carbon dioxide, methane, nitrous oxide, ozone, and fluorocarbons.

Heavy crude oil
Dark oil of high density. It flows slowly and must often be diluted before it can travel via pipeline. It is more expensive to refine than light crude.

Hydrocarbon
A compound substance containing only hydrogen and carbon.

In-situ oil production
Where open pit mining of the oil sands is impossible, the bitumen

must be extracted after it has been heated at the site. The heating is usually done by steam injection, and allows the bitumen to flow to the surface.

Kyoto Protocol

A protocol to the United Nations Framework Convention on Climate Change designed to commit industrialized countries to reducing their emissions of greenhouse gases associated with global warming. It was agreed in Kyoto, Japan in December 1997 and went into effect legally on 16 February 2005 following its ratification.

Lease condensate

A natural gas liquid that is recovered at the lease site in field separation facilities.

Light crude oil

Oil of low density with a low wax content that flows easily and is relatively cheap to refine.

Natural gas

A gas consisting mainly of methane that is used as a fuel for heating, electric power generation, and household appliances.

Oil sands (or tar sands)

Loose sand or sandstone that contains a tar-like bitumen, which can be mined and processed into "synthetic" oil. The largest known deposits are in northern Alberta.

OPEC

The Organization of Petroleum Exporting Countries was formed in 1961 to control the volume of oil production and thus to influence its price. The member countries are Algeria, Angola, Ecuador, Iran, Iraq, Kuwait, Libya, Nigeria, Qatar, Saudi Arabia, the United Arab Emirates, and Venezuela.

Peak oil

For any given area, this term refers to the point when the maximum production of oil takes place. As production inevitably declines, the price of oil is expected to rise.

Primary energy

The energy contained in natural resources prior to processing.

Proved reserves

The estimated amount of oil or natural gas that can be extracted from a field with existing technologies and under existing economic conditions.

Renewable energy

Energy from sources (including sun, wind, rain, tides, and geothermal heat) that can in principle be replaced through natural processes.

Shale gas

Natural gas produced from shale, a common form of sedimentary rock. Usually the rock has to be fractured by horizontal drilling to release the gas.

Vertically integrated company

A company that owns or controls multiple stages of production from extraction of raw materials to final sale. For example, Suncor mines the oil sands, processes the oil, and sells it to consumers at Sunoco and Petro-Canada gas stations.

Notes

Chapter 1
[1]This summary is based on Mommer (2002), Parra (2004), Shelley (2005), Warnock (2006), and Yergin (1992).

Chapter 2
[1]This was a government-industry collaboration formed in 1968 to explore in the Canadian Arctic. The federal government was the major shareholder.

Chapter 3
[1]This box summarizes research reported by Susanne Ottenheimer (1993).
[2]Interview of key informant conducted in 2008 by the author and Sean Cadigan.

Chapter 4
[1]Challenges to the IPCC analysis and responses are available on several websites: www.realclimate.org, representing most climate scientists; www.climateaudit.org, representing key critics. Shortly before the Copenhagen meeting, hackers invaded the computing system at the University of East Anglia (UK) and released emails that appeared at first glance to discredit some conclusions of scientists who have been central to the IPCC analysis. See Mooney (2009), Walsh (2009) and various postings on realclimate.org for statements—convincing to me and many others—that leave the original analysis secure. This is not to say that data should be kept hidden away and unavailable to critics. Following the email disclosures, Wikipedia published a list of selected email statements

and responses at http://en.wikipedia.org/wiki/Climatic_Research_Unit_e-mail_hacking_incident.

[2]Those countries classified as developed nations and nations with economies in transition that appear in Annex 1 of the Convention.

Chapter 6

[1]On 14 August 2003, a massive failure of the interconnected power grids in large parts of the northeast US and Ontario cut power to millions of homes and businesses. Most of Ontario was affected. The government declared a state of emergency for two days and all consumers were asked to minimize their use of electricity as supply was gradually restored.

Further Reading

Introduction

Ommer, R., et al. 2007. *Coasts under Stress: Restructuring and Social-Ecological Health*. Montreal and Kingston: McGill-Queen's University Press.

Shelley, T. 2005. *Oil: Politics, Poverty and the Planet*. New York: Zed Books.

Chapter 1

Deffeyes, K.S. 2005. *Beyond Oil: The View from Hubbert's Peak*. New York: Hill and Wang.

Falola, T., and A. Genova. 2005. *The Politics of the Global Oil Industry: An Introduction*. Westport, CT: Praeger.

International Energy Agency (IEA). 2009. *Key World Energy Statistics 2009*. Paris: IEA. Available online.

Parra, Francisco. 2004. *Oil Politics: A Modern History of Petroleum*. London: I.B.Tauris.

Yergin, D. 1992. *The Prize: The Epic Quest for Oil, Money & Power*. New York: Free Press.

Chapter 2

Doern, G.B. 2005. *Canadian Energy Policy and the Struggle for Sustainable Development*. Toronto: University of Toronto Press.

Fossum, J.E. 1997. *Oil, the State, and Federalism: The Rise and Demise of Petro-Canada as a Statist Impulse*. Toronto: University of Toronto Press.

Laxer, G. 2008. Freezing in the Dark: Why Canada Needs Strategic Petroleum Reserves. Report to the Parkland Institute and Polaris Institute. Available online.

Laxer, J. 1983. *Oil and Gas: Ottawa, the Provinces and the Petroleum Industry*. Toronto: James Lorimer.

Chapter 3

Bankes, N., and M.M. Wenig. 2005. Northern Gas Pipeline Policy and Sustainable Development, Then. And Now? In *Canadian Energy Policy and the Struggle for Sustainable Development*, ed. G.B. Doern. Toronto: University of Toronto Press.

Cadigan, Sean T. 2010. Organizing Offshore: Labour Relations, Industrial Pluralism and Order in the Newfoundland and Labrador Oil Industry, 1997-2006. In *Work on Trial: Cases in Context*, eds. E. Tucker and J. Fudge. Toronto: Osgoode Society and Irwin.

Dunning, T. 2008. *Crude Democracy: Natural Resource Wealth and Political Regimes*. Cambridge: Cambridge University Press.

House, J.D. 1985. *The Challenge of Oil: Newfoundland's Quest for Controlled Development*. St. John's, NL: ISER.

Voutier, K., B. Dixit, P. Millman, J. Reid, and A. Sparkes. 2008. Sustainable Energy Development in Canada's Mackenzie Delta-Beaufort Sea Coastal Region. *Arctic* 61 (supp. 1):103–10.

Chapter 4

Canada. 2009. Environment Canada. A Climate Change Plan for the Purposes of the Kyoto Protocol Implementation Act—May 2009. Ottawa.

Maslin, M. 2009. *Global Warming: A Very Short Introduction*. Oxford: Oxford University Press.

Schmidt, G., and J. Wolfe. 2009. *Climate Change: Picturing the Science*. New York: W.W. Norton.

Stern, N. 2007. *The Economics of Climate Change: The Stern Review*. Cambridge: Cambridge University Press.

White, R. 2010. *Climate Change in Canada*. Don Mills, ON: Oxford University Press.

Chapter 5

Marsden, W. 2008. *Stupid to the Last Drop*. Toronto: Vintage Canada.

Pratt, L. 1976. *The Tar Sands: Syncrude and the Politics of Oil*. Edmonton: Hurtig.

Söderbergh, B., F. Robelius, and K. Aleklett. 2007. A Crash Program Scenario for the Canadian Oil Sands Industry. *Energy Policy* 35: 1931–47.

Woynillowicz, D., and C. Severson-Baker. 2006. Down to the Last Drop? The Athabasca River and Oil Sands. Edmonton: Pembina Institute. Available online.

Chapter 6
Beber, Sharon. 2003. *Power Play: the Fight to Control the World's Electricity*. New York: New Press.
Cohen, M.G. 2006. Electricity Restructuring's Dirty Secret: The Environment. In *Nature's Revenge: Reclaiming Sustainability in an Age of Corporate Globalization*, eds. J. Johnston, M. Gismondi, and J. Goodman. 73–95. Peterborough, ON: Broadview Press.
Soderholm, P. 2008. The Political Economy of International Green Certificate Markets. *Energy Policy* 36:2051–2062.
Swift, J., and K. Stewart. 2004. *Hydro: the Decline and Fall of Ontario's Electric Empire*. Toronto: Between the Lines.

Chapter 7
Bradford, T. 2006. *The Solar Revolution: the Economic Transformation of the Global Solar Energy Industry*. Cambridge, MA: MIT Press.
Breukers, S., and M. Wolsink. 2007. Wind Power Implementation in Changing Institutional Landscapes: An International Comparison. *Energy Policy* 35: 2737–50.
Haggett, Claire. 2008. Over the Sea and Far Away? A Consideration of the Planning, Politics and Public Perception of Offshore Wind Farms. *Journal of Environmental Policy and Planning* 10:289–306.
Verbruggen, A. 2008. Renewable and Nuclear Power: a Common Future? *Energy Policy* 36:4036–4047.

Chapter 8
Dennis, K., and J. Urry. 2008. *After the Car*. Cambridge: Polity.

References

Abele, F. 2005. The Smartest Steward? Indigenous People and Petroleum-based Economic Development in Canada's North. In *Canadian Energy Policy and the Struggle for Sustainable Development*, ed. G.B. Doern. 223–45. Toronto: University of Toronto Press.

Adam, D. 2008. Exxon to Cut Funding to Climate Change Denial Groups. *The Guardian*, 28 May.

———. 2009. World Will not Meet 2°C Warming Target, Climate Change Experts Agree. *The Guardian*, 14 April.

Alberta. Oil Sands Ministerial Strategy Committee. 2006. Investing in our Future: Responding to the Rapid Growth of Oil Sands Development—Final Report. Edmonton. Available online.

———. 2008. Department of Energy. Launching Alberta's Energy Future: Provincial Energy Strategy. Edmonton. Available online.

———. 2009a. Department of Energy. Facts and Statistics. Edmonton. Available online.

———. 2009b. Department of Aboriginal Relations. Facts about Aboriginal People in Alberta. Edmonton. Available online.

Aleklett, K., and C. J. Campbell. 2003. The Peak and Decline of World Oil and Gas Production. *Minerals and Energy* 18: 5–20.

Alvarez, P.R., M. Cleland, and R. Gibbins. 2008. National Energy Security from an Exporter's Perspective: The Canadian Experience. Paper presented at the North Pacific Energy Security Conference, Honolulu.

Angus Reid. 2010. Premiers Williams and Wall Still Popular; Graham Now at the Bottom. Angus Reid Public Opinion. Available online.

Association for the Study of Peak Oil (ASPO). 2006. Newsletter 63 (March). Available online.

Bailey, C., P. Sinclair, and M. DuBois. 2003. Pushing Paper: Market Demand and Market-based Environmentalism. Paper presented at the Rural Sociological Society, Montreal.

Bailey, R. 2006. Peak Oil Panic. *Reason Online.* May. Available online.

Baird, M. 2002. Hebron Shelved—For Now. *The Telegram* (St. John's, NL), 14 February.

———. 2006. Negotiate, NOIA Says. *The Telegram* (St. John's, NL), 22 June.

Bankes, N., and M.M. Wenig. 2005. Northern Gas Pipeline Policy and Sustainable Development, Then. And Now? In *Canadian Energy Policy and the Struggle for Sustainable Development*, ed. G.B. Doern. 246–71. Toronto: University of Toronto Press.

Barton, B. 2004. *Energy Security: Managing Risk in a Dynamic Legal and Regulatory Environment.* Oxford: Oxford University Press.

BC Hydro. 2009. History. Available online.

Beaufort Sea Partnership (BSP). 2009. Integrated Management in the Beaufort Sea: A Brief Overview. Available online.

Beaufort Sea Strategic Regional Committee (BSSR Committee). 2008. Beaufort Sea Strategic Regional Plan of Action. Available online.

Beber, S. 2003. *Power Play: The Fight to Control the World's Electricity.* New York: New Press.

Berger, T.R. 1988. *Northern Frontier, Northern Homeland: The Report of the Mackenzie Valley Pipeline Inquiry.* Vancouver: Douglas & McIntyre.

Bergin, T. 2006. Total Sees Peak Oil Output around 2020. *Energy Bulletin,* 6 June.

Bergman, B. 2002. Kyoto Accord Opposition Growing. *Maclean's,* 14 October.

Bielo, D. 2009. Atomic Weight: Balancing the Risks and Rewards of a Power Source. *Scientific American,* 29 January.

Black, E. 2006. *Internal Combustion: How Corporations and Governments Addicted the World to Oil and Derailed the Alternatives.* New York: St. Martin's Press.

Boxell, J. 2004. Top Oil Groups Fail to Recoup Exploration Costs. *New York Times,* 9 October.

BP. 2004. Energy in Focus: Statistical Review of World Energy 2003. Available online.

———. 2006. It's Time to Go on a Low Carbon Diet. Available online.

———. 2007. Annual Report and Accounts 2007. Available online.

———. 2009. Statistical Review of World Energy 2008. Available online.

Bramley, M., P. Sadik, and D. Marshall. 2009. Climate Leadership, Economic Prosperity. Drayton Valley, AB / Vancouver: Pembina Institute / David Suzuki Foundation.

Brownsey, K. 2005. The Best of Times? Petroleum Politics in Canada. Paper presented at annual meeting of the Canadian Political Science Association. Halifax, May.

Bregha, F. 2009. National Energy Program. *The Canadian Encyclopedia.* Available online.

Breukers, S., and M. Wolsink. 2007. Wind Power Implementation in Changing Institutional Landscapes: An International Comparison. *Energy Policy* 35: 2737–50.

British Columbia. 2007. Department of Energy, Mines, and Petroleum Resources. The BC Energy Plan: A Vision for Clean Energy Leadership. Victoria. Available online.

———. 2009. Department of Energy, Mines, and Petroleum Resources. BC Energy Plan: Report on Progress. Victoria. Available online.

Brown, L.R. 2009. *Plan B: 4.0 Mobilizing to Save Civilization*. New York: W.W. Norton.

Bunn, M. 2008. Expanding Nuclear Energy, Preventing Nuclear Terrorism. Paper presented to Energy and Security Search Seminar, Cambridge, MA. Available online.

Campbell, C.J. 2005. *Oil Crisis*. Brentwood, UK: Multi-Science.

Canada. 1985. Atlantic Accord: Memorandum of Agreement between the Government of Canada and the Government of Newfoundland and Labrador on Offshore Oil and Gas Resource Management and Revenue Sharing. Ottawa. Available online.

———. 2007. Environment Canada. A Climate Change Plan for the Purposes of the Kyoto Protocol Implementation Act. Ottawa. Available online.

———. 2008a. Environment Canada. Canada's Greenhouse Gas Emissions: Understanding the Trends, 1990–2006. Ottawa.

———. 2008b. Natural Resources Canada. Alternative Fuels in Canada: Making Choices Today for a Better Tomorrow. Ottawa. Available online.

———. 2009a. Department of Fisheries and Oceans Canada. The Beaufort Sea Integrated Planning Atlas. Available online.

———. 2009b. Department of Foreign Affairs and International Trade Canada. Energy Security: A Canadian Perspective. Ottawa. Available online.

———. 2009c. Environment Canada. A Climate Change Plan for the Purposes of the Kyoto Protocol Implementation Act. Ottawa. Available online.

———. 2009d. Natural Resources Canada. The Canadian Nuclear Industry and its Economic Contributions. Ottawa. Available online.

———. 2009e. Natural Resources Canada. Overview of Canada's Energy Policy. Ottawa. Available online.

Canada-United States Free Trade Agreement. 1988. Available online.

Canadian Association of Petroleum Producers (CAPP). 2009. Environmental Aspects of Oil Sands Development. Available online.

Canadian Press. 2009. Northern Businesses Wait in Frustration for Mackenzie Valley Pipeline Decision. 18 June. Available online.

Canadian Wind Energy Association (CanWEA). 2008. National Poll Finds that Canadians Prefer Wind Energy over All Other Sources.

Poll conducted by the Strategic Counsel for CanWEA. Available online.

———. 2009. Wind Farms. Available online.

Castaldo, J. 2008. Canada in 2020—Alternative Energy: Out of Juice? *Canadian Business*, 27 October.

Cattaneo, C. 2007. Why Did Big Oil Cave on Hebron? Hibernia South Played Key Role in Deal Going Forward. *Financial Post*, August 28.

———. 2009. Mackenzie Valley Pipeline "A National Embarrassment": MGM's Sykes. *National Post*, 5 May.

CBC News. 2002. Alberta Launches Campaign against Kyoto. 18 September. Available online.

———. 2006. Williams Chafed by PM's Stance on Oil Standoff. 8 September. Available online.

———. 2007a. Harper Denies Claims in N.L. Attack Ads. 28 May. Available online.

———. 2007b. The MacKenzie Valley Pipeline. 31 December. Available online.

———. 2007c. Hydro Quebec Shelves Smart Meters. 10 December. Available online.

———. 2008. Carbon Tax Proposal a Non-starter in Alberta. 8 January. Available online.

———. 2009a. Canada Part of Copenhagen Climate Deal. 18 December. Available online.

———. 2009b. Global Climate Change Treaty Unlikely: Harper Urges Climate Rules for all Countries. 14 November. Available online.

———. 2009c. 21 Charged in Greenpeace Oilsands Protest. 1 October. Available online.

———. 2009d. Alberta First Nations Place Anti-oilsands Ad in Major U.S. Paper. 17 February. Available online.

———. 2009e. Green Groups Begin Campaign to Tell Obama about Alberta's "Dirtiest Oil." 10 February. Available online.

CERES. 2004. Global Warming Resolutions at U.S. Oil Companies Bring Policy Commitments from Leaders, and Record High Votes at Laggards.

Chevron. 2006. Chevron Announces Plans to Suspend Hebron Activities. 3 April. Available online.

———. 2007. Annual Report. Available online.

Choynowski, P. 2004. Restructuring and Regulatory Reform in the Power Sector: Review of Experience and Issues. Manilla, Philippines: Asian Development Bank.

Cohen, M.G. 2002. From Public Good to Private Exploitation: Electricity Deregulation, Privatization and Continental Integration. Halifax: Canadian Centre for Policy Alternatives. Available online.

———. 2006. Electricity Restructuring's Dirty Secret: The Environment. In *Nature's Revenge: Reclaiming Sustainability in an Age of Corporate*

Globalization, eds. J. Johnston, M. Gismondi, and J. Goodman. 73–95. Peterborough, ON: Broadview Press.

ConocoPhillips. 2007. Annual Report. Available online.

———. 2008. First Quarter Report 2008. Available online.

Courchene, T.J. 2005. Energy Prices, Equalization, and Federalism. *Policy Options*, 40–45.

CTV News. 2007a. N.L. Premier Announces Deal on Hebron Oil Project. 22 August. Available online.

———. 2007b. The Atlantic Accord Dispute: Questions and Answers. 11 June. Available online.

———. 2007c. Most Willing to Sacrifice for Environment: Poll. 26 January. Available online.

———. 2009. Newfoundland Liberal MP Vows to Vote against Budget. 30 January. Available online.

Curry, B., and D. Walton. 2009. Climate Change Report "Irresponsible," Prentice Says. *Globe and Mail*, 29 October.

Davidson, P. 2007. Shocking Electricity Prices Follow Deregulation. *USA Today*, 9 August.

Deffeyes, K.S. 2005. *Beyond Oil: The View from Hubbert's Peak*. New York: Hill and Wang.

Dewees, D.N. 2005. Electricity Restructuring in Canada. In *Canadian Energy Policy and the Struggle for Sustainable Development*, ed. G.B. Doern. 128–50. Toronto: University of Toronto Press.

Dietz, T., A. Dan, and R. Shwom. 2007. Support for Climate Change Policy: Social Psychological and Social Structural Influences. *Rural Sociology* 72: 185–214.

Doern, G.B. 2005. *Canadian Energy Policy and the Struggle for Sustainable Development*. Toronto: University of Toronto Press.

Doern, G.B., and M. Gattinger. 2003. *Power Switch: Energy Regulatory Governance in the Twenty-First Century*. Toronto: University of Toronto Press.

Doggett, T. 2009. EPA to Review Emission Rules for Power Plants. Reuters, 27 April. Available online.

Dorn, J.G. 2008. World Geothermal Power Generation Nearing Eruption. Earth Policy Institute. Available online.

DuBois, G.L. 2009. Accident Casts Fresh Doubt on Nuclear Safety. *Baltimore Sun*, 25 November.

The Economist. 2009. Nuclear Contamination. 19 November.

Energy Information Administration (EIA). 2008. International Energy Annual 2006. 8 December. Available online.

———. 2009. World Proved Reserves of Oil and Natural Gas, Most Recent Estimates. Available online.

EKOS. 2002. Public Attitudes toward the Kyoto Protocol. Available online.

European Commission. 2008. Sweden: Renewable Energy Fact Sheet. Directorate-General for Energy and Transport. Available online.

European Renewable Energy Council (EREC) and Greenpeace. 2008. Energy Revolution: A Sustainable Global Energy Outlook. Available online.

Evans-Pritchard, A. 2009. Energy Crisis is Postponed as New Gas Rescues the World. *The Telegraph*, 11 October.

ExxonMobil. 2006a. 2005 Summary Annual Report. Available online.

———. 2006b. Tomorrow's Energy: A Perspective on Energy Trends, Greenhouse Gas Emission and Future Energy Options.

———. 2008. ExxonMobil Taking on the World's Toughest Energy Challenges: 2008 Corporate Citizenship Report.

———. 2009. 2008 Financial and Operating Review.

Firestone, J., and W. Kempton. 2007. Public Opinion about Large Offshore Wind Power: Underlying Factors. *Energy Policy* 35: 1584–98.

Follath, E., and A. Jung. 2006. Interview with BP CEO Lord Browne. We Take the Problem of Climate Change Seriously. *Der Spiegel*, 12 June. Available online.

Fossum, J.E. 1997. *Oil, the State, and Federalism: The Rise and Demise of Petro-Canada as a Statist Impulse*. Toronto: University of Toronto Press.

Foster, J.B. 2002. *Ecology against Capitalism*. New York: Monthly Review Press.

Froschauer, K. 1999. *White Gold: Hydroelectric Power in Canada*. Vancouver: University of British Columbia Press.

Gattinger, G., B. Doern, and M. Gattinger. 2003. *Power Switch: Energy Regulatory Governance in the Twenty-First Century*. Toronto: University of Toronto Press.

Global Wind Energy Council. 2010. Wind is a Global Power Source. Available online.

Grant, J., S. Dyer, and D. Woynillowicz. 2009. Clearing the Air on Oil Sands Myths. Edmonton: Pembina Institute. Available online.

Gray, J. 2008. That's Danny Billions to You. *Globe and Mail*, 25 January.

Green Energy Act Alliance. 2009. Poll Shows Overwhelming Support for Ontario Green Energy Act. Available online.

Hadekel, P. 2009. Gas Prices Chill Sector's Future Here. *Montreal Gazette*, 9 December.

Hausman, W.J., P. Hertner, and M. Wilkins. 2008. *Global Electrification: Multinational Enterprise and International Finance in the History of Light and Power 1878–2007*. Cambridge: Cambridge University Press.

Heinberg, R. 2005. *The Party's Over: Oil, War and the Fate of Industrial Societies*. Gabriola Island, BC: New Society Publishers.

Henton, D. 2008. First Nations Unite to Fight the Tar Sands. *Edmonton Journal*, 19 August.

Hirsch, R.L., R. Bezdek, and R. Wendling. 2005. Peaking of World Oil

Production: Impacts, Mitigation, and Risk Management. Report to US Department of Energy. Available online.

Hornig, J.F., ed. 1999. *Social and Environmental Impacts of the James Bay Hydroelectric Project*. Montreal: McGill-Queen's University Press.

Hotten, R. 2008. BP Mulls Sale of its Green Energy Division. *The Telegraph*, 28 February.

House, J.D. 1985. *The Challenge of Oil: Newfoundland's Quest for Controlled Development*. St. John's, NL: ISER.

Howell, K. 2009. Some See Exxon Investments Into Alternative Energy Signaling "Paradigm Shift" for Big Oil. *New York Times*, 16 July. Available online.

Hydro-Québec. 2009. Powering Our Future: Annual Report 2008. Montreal: Hydro-Québec.

Institute for Energy and Environmental Research (IEER). 2005. Uranium: Its Uses and Hazards. Available online.

Intergovernmental Panel on Climate Change (IPCC). 2007. Summary for Policymakers. *Climate Change 2007: Mitigation. Contribution of Working Group III to the Fourth Assessment Report of the Intergovernmental Panel on Climate Change*, ed. B. Metz, O.R. Davidson, P.R. Bosch, R. Dave, and L.A. Meyer. Cambridge: Cambridge University Press.

International Energy Agency (IEA). 2009. *Key World Energy Statistics 2009*. Paris: IEA. Available online.

International Petroleum Monthly. 2009. US Energy Information Administration. May. Available online.

Ipsos-Reid. 2002. Albertans and the Kyoto Protocol: A Quantitative Survey. Available online.

Jackson, C. 2006. Door always Open. Room for Future Talks on Hebron-Ben Nevis: Premier. *The Telegram* (St. John's, NL), 11 April.

Joberta, A., P. Laborgneb, and S. Mimlerb. 2007. Local Acceptance of Wind Energy: Factors of Success Identified in French and German Case Studies. *Energy Policy* 35: 2751–60.

Jones, J. 2009. Economics, Politics Chill Mackenzie Pipeline Dreams after Decades. *Globe and Mail*, 9 December.

Joskow, P.L. 2001. California's Electricity Crisis. *Oxford Review of Economic Policy* 17:365–388.

Jubak, J. 2008. Is ExxonMobil's Future Running Dry? *MSN Money*. Available online.

Kerr, A. 2007. Serendipity is Not a Strategy: The Impact of National Climate Programs on Greenhouse Gas Emissions. *Area* 39:418–30.

Lalond, M. 2009. Want $60 for Your Old Fridge? Call Hydro Quebec. *Montreal Gazette*, 13 May.

Lampton, C. 2009. What Are the Benefits of Hydrogen-powered Vehicles? http://www.howstuffworks.com.

Laxer, G. 2008. Freezing in the Dark: Why Canada Needs Strategic

Petroleum Reserves. Report to the Parkland Institute and Polaris Institute. http://www.ualberta.ca/parkland.

Laxer, J. 1983. *Oil and Gas: Ottawa, the Provinces and the Petroleum Industry.* Toronto: James Lorimer.

Lessard, D. 2009. Privatisation d'Hydro: "Pas dans les cartons du gouvernement." *La Presse*, 4 February.

Levine, S. 2009. Is ExxonMobil Heading for a Fall? *MSN Money*, 2 February. Available online.

Lougheed, P. 2006. Interview with Peter Lougheed: Sounding an Alarm for Alberta. *Policy Options*, 5–9 September.

Lundeberg, S. 2009. Challenges of the Scandinavian Countries. Swedish Environmental Protection Agency. Paper presented at Bio-Waste: Need for EU-Legislation? Brussels.

Macalister, T. 2009. BP and Shell Warned to Halt Campaign Against US Climate Change Bill. *Guardian*, 19 August. Available online.

MacDonald, J.A.C. 2008. Boomtown Hustles to Keep Up with Breakneck Growth. *Edmonton Journal*, 26 March. Available online.

MacDonald, M. 2001. Newfoundland Government Approves White Rose Offshore Oil Project. Canadian Press Newswire. Toronto.

Madslien, J. 2009. All Change as Gas Reserves Soar. BBC News. Available online.

Marsden, W. 2009. Big Oil's Relentless Lobby. *Montreal Gazette*, 4 December.

Maslin, M. 2009. *Global Warming: A Very Short Introduction*. Oxford: Oxford University Press.

Mathews, J.A. 2007. Biofuels: What a Biopact between North and South Could Achieve. *Energy Policy* 35: 3550–70.

McCarthy, S. 2009a. In Mackenzie Valley, Frustration and a Sense of Foreboding. *Globe and Mail*, 30 August.

———. 2009b. Copenhagen Fell Victim to a World Divided. *Globe and Mail*, 20 December.

McCright, A.M., and R.E. Dunlap. 2003. Defeating Kyoto: The Conservative Movement's Impact on U.S. Climate Change Policy. *Social Problems* 50: 348–73.

McIntyre, S. 2008. How Do We Know that 1998 Was the Warmest Year of the Millennium? Paper presented at Ohio State University, Columbus, OH. Available online.

Mommer, B. 2002. *Global Oil and the Nation State*. Oxford: Oxford University Press.

Mooney, C. 2009. Not So Swift, Hackers: Why the Scandal Sometimes Called ClimateGate is Overblown. Available online.

Mouawad, J. 2009a. Estimate Places Natural Gas Reserves 35 Percent Higher. *New York Times*, 17 June.

———. 2009b. Oil Giants Loath to Follow Obama's Green Lead. *New York Times*, 8 April.

Mufson, S., and J. Eilperin. 2006. Energy Firms Come to Terms With Climate Change. *Washington Post*, 24 November.

National Energy Board of Canada (NEB). 2008. Canada's Oil Sands: Opportunities and Challenges to 2015: Questions and Answers. Ottawa: National Energy Board of Canada. Available online.

———. 2009. Energy Brief: Understanding Canadian Shale Gas. Ottawa: National Energy Board of Canada. Available online.

Nation Talk. 2009. Suncor and Fort McKay First Nation Celebrate Business Incubator Opening. 3 June. Available online.

Newfoundland and Labrador. 2003. Department of Mines and Energy. Minister Noel Says Government Succeeding in Increasing Offshore Benefits. Press release, 19 June. St. John's.

———. 2005. Atlantic Accord. Available online.

———. 2008. The Economy 08: Securing a Sustainable Future. Available online.

———. n.d. Community Accounts. Online statistics provided by the Newfoundland and Labrador Statistics Agency.

Newnet. 2009. Renewable Energy Developer RES Canada and Energy Transportation Company Enbridge Agree to Develop 99MW Canadian Wind Project. 20 November. Available online.

Nuclear Age Peace Foundation. 2009. What is Nuclear Fission? Available online.

Nunavut. 2009. Department of Economic Development & Transportation. Economic Development, Oil and Gas. Iqaluit. Available online.

Oil Drum. 2009. Peak Oil Update—July 2009: Production Forecasts and EIA Oil Production Numbers. 7 July. Available online.

Olofsson, L. 2005. Swedish Government Embraces Peak Oil and Looks towards Biofuels. *Energy Bulletin*, 16 December.

Ommer, R., et al. 2007. *Coasts under Stress: Restructuring and Social-Ecological Health.* Montreal and Kingston: McGill-Queen's University Press.

Ontario. 2002. Ontario Electricity Pricing, Conservation and Supply Act, 2002. Toronto. Available online.

———. 2009. Green Energy and Green Economy Act. Toronto. Available online.

Ontario Energy Board (OEB). 2009. Integrated Power System Plan Review. Toronto: Queen's Printer.

Ontario Power Authority (OPA). 2005. Supply Mix Advice and Recommendations: Report to the Ontario Ministry of Energy. Available online.

Orange, R. 2004. Oil Supply to Peak Sooner than We Think, Says BP Scientist. *The Business*, 7 November. Available online.

Ottenheimer, S. 1993. Fish and Oil Don't Mix: Power Relations at the Bull Arm Construction Site, Trinity Bay, Newfoundland. MA thesis. Memorial University, St. John's, Canada.

Parra, Francisco. 2004. *Oil Politics: A Modern History of Petroleum*. London: I.B.Tauris.

Pembina Institute. 2007. Sources of Renewable Energy. Available online.

——. 2009. New Federal Numbers Confirm the End of Support for Renewable Power. Press Release, 10 December. Available online.

Plourde, A. 1991. The NEP Meets the FTA. *Canadian Public Policy* 17: 14–24.

Pratt, L. 1976. *The Tar Sands: Syncrude and the Politics of Oil*. Edmonton: Hurtig.

Reguly, E. 2007. Jim Buckee: Talisman's Retired Contrarian Picks His Next Fight. *Globe and Mail*, 8 October.

Rennie, S. 2009a. Canada's Kyoto View Triggers a Walkout. *The Toronto Star*, 13 October.

——. 2009b. Harper Says Climate Plan Won't Change Much. *Toronto Sun*, 29 November.

Robelius, F. 2007. Giant Oil Fields—The Highway to Oil: Giant Oil Fields and Their Importance for Future Oil Production. PhD. Diss. Uppsala University, Uppsala, Sweden.

Rutledge, I. 2005. *Addicted to Oil: America's Relentless Drive for Energy Security*. London: I.B. Tauris.

Samuel, H. 2009. EDF "Sends Used Nuclear Material" to Siberia. *The Telegraph*, 13 October.

Shell. 2007. Final Report. Available online.

——. 2008. First Quarter Report. Available online.

Shelley, T. 2005. *Oil: Politics, Poverty and the Planet*. New York: Zed Books.

Sheppard, K. 2010. Oil Companies Abandon Climate Partnership. *MotherJones*, 16 February. Available online.

Simmons, M.R. 2007. Another Nail in the Coffin of the Case against Peak Oil. *World Energy Magazine* 11 (1).

Söderbergh, B., F. Robelius, and K. Aleklett. 2007. A Crash Program Scenario for the Canadian Oil Sands Industry. *Energy Policy* 35: 1931–47.

Stern, N. 2007. *The Economics of Climate Change: The Stern Review*. Cambridge: Cambridge University Press.

Stoett, P. 2008. Canada, Kyoto, and the Conservatives. In *Encyclopedia of Earth*, ed. J. Cleveland Cutler. Washington, DC: Environmental Information Coalition.

Swift, J., and K. Stewart. 2004. Hydro: The Decline and Fall of Ontario's Electric Empire. Toronto: Between the Lines.

Taylor, P.L. 2005. In the Market but not of It: Fair Trade Coffee and Forest Stewardship Council Certification as Market-Based Social Change. *World Development* 33: 129–47.

The Telegram (St. John's, NL). 2008. The Dark Side of Oil Prosperity. 27 August.

Thomson, G. 2009. Climate-Change Deniers Horrified by Pembina Report Should Read It. *Edmonton Journal*, 31 October.

Tillerson, R. 2008. Rex Tillerson—CEO Exxon Mobil Denying Peak Oil. CNBC interview, 30 June. Available on YouTube.

Total SA. 2008. First Quarter Results. Available online.

United Nations Framework Convention on Climate Change (UNFCCC). 2009a. Greenhouse Gas Inventory Data. Available online.

———. 2009b. Kyoto Protocol Status of Ratification. Available online.

———. 2009c. Text of the Kyoto Protocol to the United Nations Framework Convention on Climate Change. Available online.

United States. 2009. Department of Energy. Ethanol Production. Available online.

United States Nuclear Regulatory Commission (US NRC). 2003. Fact Sheet on Plutonium. Available online.

Urry, J. 2008. Governance, Flows, and the End of the Car System? *Global Environmental Change* 18: 343–49.

VanderKlippe, N. 2010. Inuvik Looks Ahead to "Happy Day." *Globe and Mail*, 1 January.

van der Meer, J. 2006. Setting the Energy Scene. Speech Delivered at International Petroleum Week, London. Available online.

Voutier, K., B. Dixit, P. Millman, J. Reid, and A. Sparkes. 2008. Sustainable Energy Development in Canada's Mackenzie Delta-Beaufort Sea Coastal Region. *Arctic* 61 (supp. 1): 103–10.

Wallerstein, I. 2004. *World-Systems Analysis: An Introduction*. Durham, NC: Duke University Press.

Walsh, B. 2009. Has "Climategate" Been Overblown? *Time with CNN*, 28 December. Available online.

Warnock, J. W. 2006. *Selling the Family Silver: Oil and Gas Royalties, Corporate Profits, and the Disregarded Public*. Regina, SK: Parkland Institute and Canadian Centre for Policy Alternatives.

Williams, D. 2005. Signing of Offshore Revenue Agreement. Speech delivered on 14 February. Available online.

World Nuclear Association. 2009. Supply of Uranium. Available online.

World Wind Energy Association (WWEA). 2009. World Wind Energy Report 2008. Bonn, Germany: World Wind Energy Association.

———. 2010. World Wind Energy Report 2009. Bonn, Germany: World Wind Energy Association.

Woynillowicz, D., and C. Severson-Baker. 2006. Down to the Last Drop? The Athabasca River and Oil Sands. Edmonton: Pembina Institute.

Yergin, D. 1992. *The Prize: The Epic Quest for Oil, Money & Power*. New York: Free Press.

Zabjek, A. 2009. Too Late to Object to Oilsands Project: Province. *Edmonton Journal*, 2 September.

Index